The Dragonflies of Central America exclusive of Mexico and the West Indies.

A Guide to their Identification.

by

STEFFEN FÖRSTER

2ND EDITION

© GUNNAR REHFELDT
BRAUNSCHWEIG 2001

ODONATOLOGICAL MONOGRAPHS 2

The dragonflies of Central America excl. Mexico and the West Indies. A Guide to their Identification.

2nd edition - January 2001
First published - May 1999

Published by
Gunnar Rehfeldt, Roseggerweg 41, D-38304 Wolfenbüttel, Germany

Available from:
Steffen Förster, Kastanienallee 40a, D-38104 Braunschweig, Germany
Email: sloth92@yahoo.com , steffen-foerster@gmx.de

ISBN 3-9804366-1-6
ISSN 1434-1123

Printed by:
Wolfram Schmidt, Buchbinderei & Druckerei, Hamburger Strasse 267,
D-38114 Braunschweig, Germany

Cover photograph:

Gynacantha tibiata, male (© S. Förster), La Selva Biological Station (OTS), Costa Rica

Preface

Since the famous and outstanding work of CALVERT (1908) there hasn´t been any comparable treatment of the Central American dragonfly fauna. Especially for someone who wants to start working with the dragonflies of this region it can be difficult to accumulate all the literature needed for a proper identification. Even if one has finally collected all available information, there is still the problem of handling all those hundreds of papers, species descriptions, and different taxonomic nomenclatures used by different authors, especially when going out to the field. These problems I was forced to solve when I prepared a research project on the behavioral ecology of Central American dragonflies several years ago. Trying to find a solution, I ended up with a rather rough collection of xerox copies of identification keys from the literature. It turned out to be a very useful, almost indispensable tool not only in the field, but also at home when checking and identifying collected material.

Soon I started to think about making this collection of literature available to people who don´t have the time or the opportunities to gather all the information by themselves. In 1997, I began to revise my compilation by turning the more or less provisionally copied sheets of paper into computerized files for a more comfortable editing and reading, by adding figures to make the descriptions of morphological characters easier to understand, and by updating the original keys with new species described since the last revisional work, which often was that of CALVERT (1908).

With this collection of up to date identification keys I hope to provide an useful tool for everyone who is interested in Neotropical Odonata, especially for those who spend much time collecting and studying dragonflies in the field, or who are looking for an easy to handle tool to identify species in collections.

Preface to the 2nd edition

Because of the great interest shown in the first edition of this compilation it took a shorter time than expected for the book to be sold out, bringing up the issue of how to supply additional copies to satisfy the existing demand. Instead of simply reprinting the book, I decided to delay the second printing a little bit in order to make some improvements to the first version. Necessary changes consisted mainly of correcting typing errors, and adding and editing of a few figures. For valuable suggestions and corrections I am in debt to CARLOS ESQUIVEL and DAVID WAGNER.

Saying this, I want to encourage everyone who is using this compilation, either in the field with living specimens, or in collections, to contact me if mistakes, inconsistencies, or any other problems while using the keys are being discovered. The large number of species, and the still incomplete systematic treatment of many genera, especially regarding females, will very likely lead to major changes in the structure of many keys in the future, and all help is needed to improve following editions.

Steffen Förster
Braunschweig, 10th January 2001

Contents

Plates

Introductory comments

I want to state clearly at the very beginning that most of the information provided in the following treatise originated from literature. However, an attempt was made to test existing keys for their usefulness for identification, and to modify them if necessary. These modifications often consisted of simplifying the descriptions as much as possible to ensure an easier understanding, realized by changes in phrasing, excluding minor details and unifying the terminology. For appreciation of the taxonomic work of the many researchers, at the end of each key the literature used is cited. To give information about the degree of modification made to the original sources, I used the following three modes of citing: 'AFTER' - when only minor changes were made without changes in the structure of the original key, 'DERIVED FROM' - when structural changes were made and/or up to 25% of the total number of species included in the key were added from species descriptions not considered in the original key, 'COMPILED FROM' - when several published keys were unified to give a new one and/or more than 25% of the total number of species included in the key were added subsequently.

While preparing this compilation, decisions had to be made about which species to include in the keys. For this purpose I mainly used the lists of Odonata of Middle America provided by PAULSON (1982, 1997). Because of practical reasons I excluded about 130 species which have been recorded so far only for Mexico or the West Indies. Doubtlessly some of these species can be expected to have the potential for a wider distribution than known up to now, but including them in this compilation would have complicated and delayed the preparation and updating of many keys. It would surely be worth the effort to include them in a future edition of this compilation.

In addition to the keys for identification, I provided information on the general appearance, ecology, and behavior of the families and most of the genera as far as available. Because still very few species have been studied in the field, most of the information about ecology and behavior must be said to be rather poor.

Unfortunately, it was not possible to include color plates and photographs in this compilation. In order to aid in getting an idea about the appearance of characteristical species, I established a web-page where I listed links to some of my own photographs as well as links to other pictures and scanned images available online. This page is accessible under the URL >http://www.tu-bs.de/~y0003753/Startpage.htm<. The guides to the dragonflies and damselflies of Florida written by DUNKLE (1989, 1990) are also of great help in this respect.

HOW TO WORK WITH THE KEYS

Always be careful to check all possible alternatives before deciding how to proceed. Please read the descriptions carefully and pay attention to any character mentioned. If there are any figures available this would be indicated in the description. Sometimes figures show different characters than given in the keys, for example in the males of the genus *Argia*, where the key is based mainly on color pattern of thorax and abdomen, but the figures show the male anal appendages only. In such cases, the figures are referred to at the beginning of a key or at the end of the description of a species.

I tried to concentrate on characters which are visible with little help by a 10x to 15x hand lens, but sometimes it is necessary to check the characters with a microscope, especially when genital morphology is concerned.

If there are any difficulties in understanding the meaning of morphological termini or of the sometimes very cryptic descriptions of morphological features, please refer to Plates I-III following this introduction. I unified all the different systems of nomenclature used by different authors, especially regarding wing venation. So there should be no question about which character to look for. For working with other literature I adopted, with little modifications, the comparison of terminologies of odonate wing venation from BORROR (1943) (Tab. 1). Last but not least I added a glossary with explanations of many terminologies used for describing specific characters.

Plate I

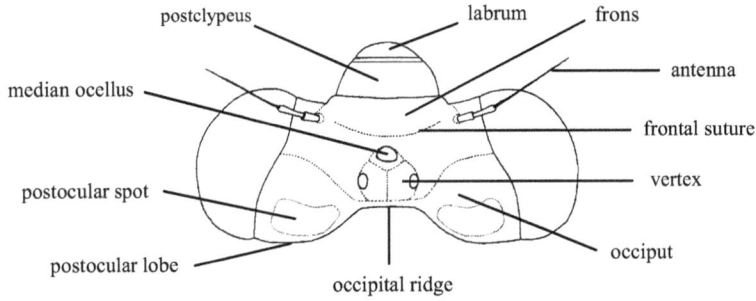

Fig. 1: Head of Zygoptera, dorsal view

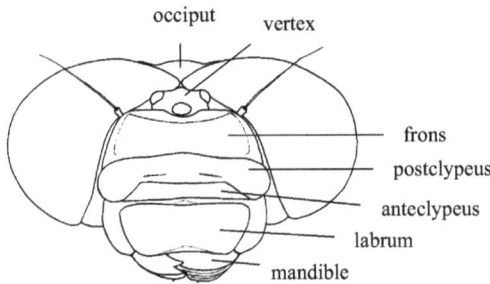

Labels: postclypeus, labrum, frons, antenna, median ocellus, frontal suture, vertex, postocular spot, occiput, postocular lobe, occipital ridge

Fig. 2: Head of Anisoptera, frontal view

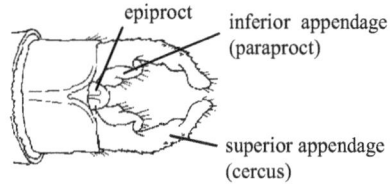

Labels: occiput, vertex, frons, postclypeus, anteclypeus, labrum, mandible

Fig. 3: Apex of male abdomen in dorsal view, Zygoptera

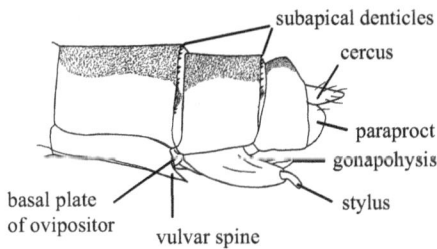

Labels: epiproct, inferior appendage (paraproct), superior appendage (cercus)

Fig. 4: Apex of female abdomen in lateral view, Zygoptera

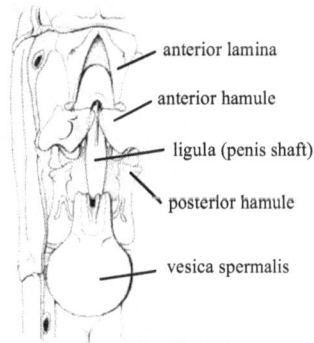

Labels: subapical denticles, cercus, paraproct, gonapophysis, stylus, basal plate of ovipositor, vulvar spine

Fig. 5: Male accessory genitalia (Calopterygidae), ventral view

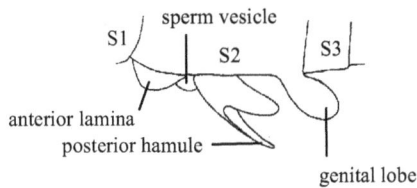

Labels: anterior lamina, anterior hamule, ligula (penis shaft), posterior hamule, vesica spermalis

Fig. 6: Male accessory genitalia (Libellulidae), lateral view

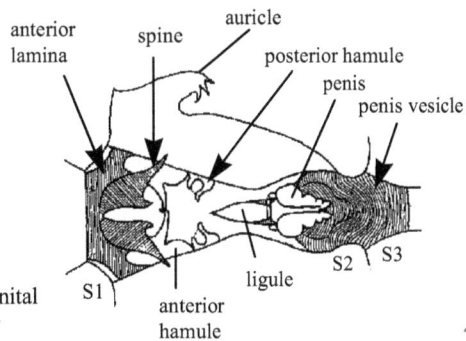

Labels: sperm vesicle, S1, S2, S3, anterior lamina, posterior hamule, genital lobe

Fig. 7: Male accessory genital (Aeshnidae), ventral view

Labels: anterior lamina, spine, auricle, posterior hamule, penis, penis vesicle, S2, S3, S1, anterior hamule, ligule

Plate II

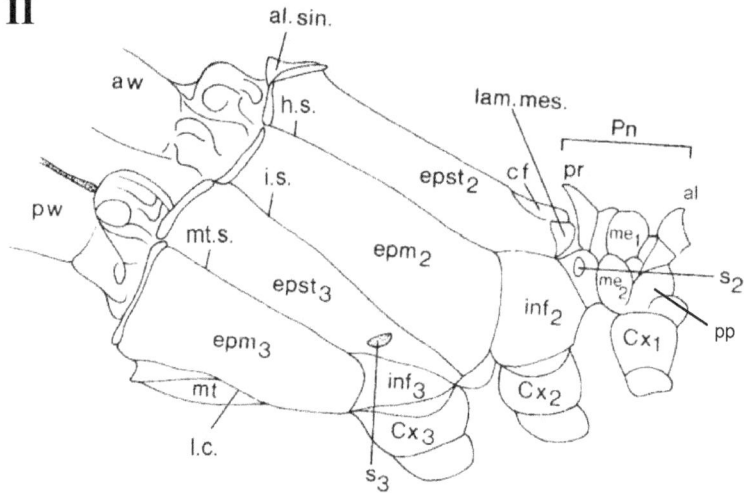

Fig. 8: Thorax of Zygoptera (Calopterygidae), lateral view

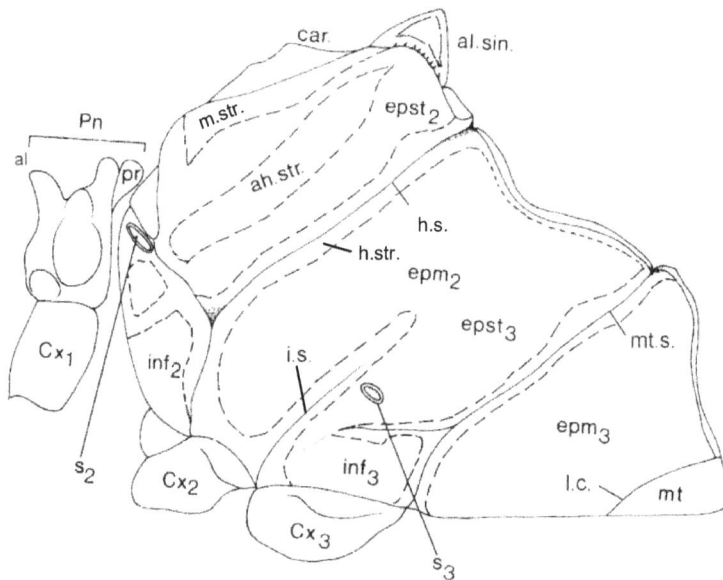

Fig. 9: Thorax of Anisoptera (Gomphidae), lateral view

Abbreviations:

al - anterior lobe of prothorax; **ah. str.** - antehumeral stripe; **al. sin.** - alar sinus; **aw** - anterior wing implant; **car** - carina; **cf** - carinal fork; Cx_1 - coxa of first pair of legs; Cx_2 - coxa of leg 2; Cx_3 - coxa of leg 3; epm_2 - mesepimeron; epm_3 - metepimeron; $epst_2$ - mesepisternum; $epst_3$ - metepisternum; **h. s.** - humeral suture (mesopleural suture); **h. str.** - humeral stripe; inf_2 - mesinfraepisternum; inf_3 - metinfraepisternum; **i.s.** - interpleural sture; **l.c.** - latero-ventral carina; **lam.mes.** - lamina mesostigmalis; **me1, me2** - median lobes of pronotum; **m. str.** - middorsal stripe; **mt** - metasternum; **mt.s.** - metapleural suture; **Pn** - pronotum; **pp** - propleuron; **pr** - posterior rim of pronotum; **pw** - posterior wing implant; s_2 - mesostigma; s_3 - metastigma;

For explanation of abbreviations see Table 1:
"Comparison of terminologies of odonate wing venation".

Plate III

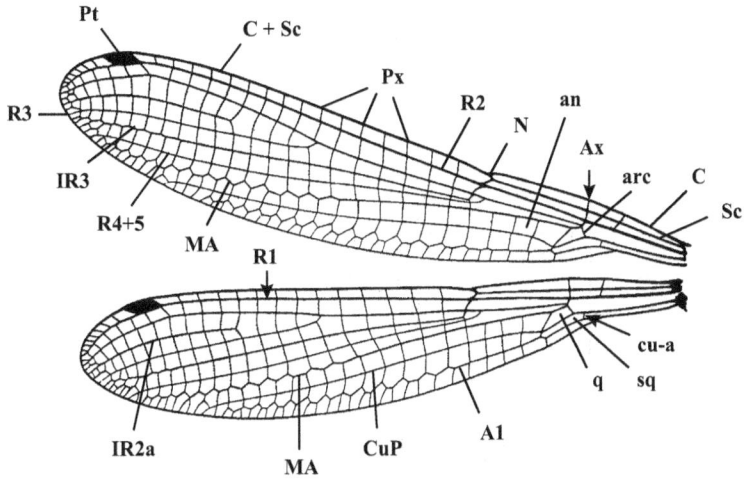

Fig. 10: Wings of Zygoptera (Coenagrionidae)

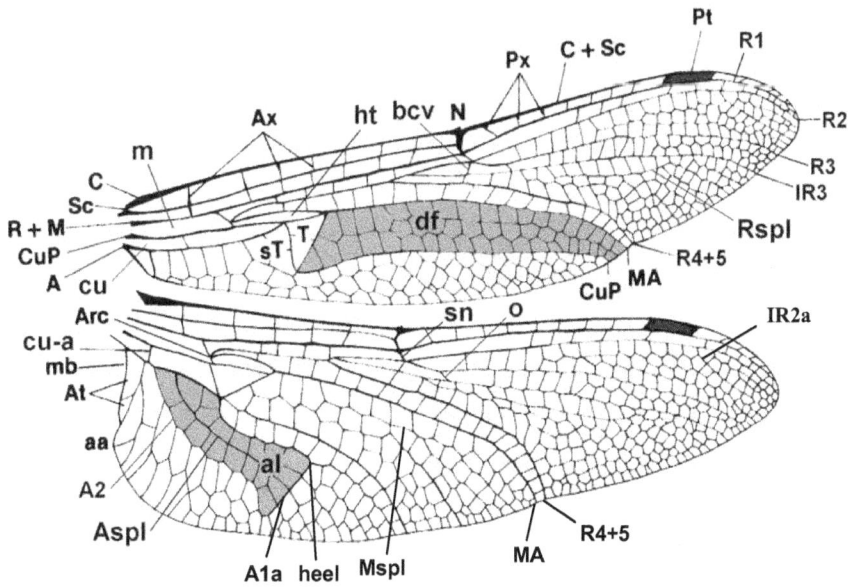

Fig. 11: Wings of Anisoptera (Corduliidae)

Tab. 1: Comparison of terminologies of odonate wing venation (After: BORROR, 1945)

Terminology used in this compilation		Other terminologies			
Name	Abbr.	Selys	Kirby	Needham	Tillyard 1917
costa	C	costa	costa	C	C
subcosta	Sc	subcosta	internodal radius	Sc	Sc
radius + media (wing-base to arc.)	R + M	median	principal radius	R + M	R + M
first radius	R1	median		R1	R
radial sector	Rs	upper sector of arculus	upper sector of arculus	M1-3	M1-3
second radius	R2	principal sector	subcostal radius	M1	M1
second radius intercalaries	IR2a	postnodal sector		M1a	
third radius	R3	nodal sector	nodal sector	M2	M2
third radius intercalary	IR3	subnodal sector	subnodal sector	Rs	Rs
radial supplement	Rspl	supplementary sector next below subnodal sector		Rspl	Rspl
fourth and fifth radius	R4+5	median sector	upper sector of arculus	M3	M3
anterior media	MA	lower sector of arculus, short sector	lower sector of arculus	M4	M4
median supplement	Mspl	supplementary sector next below short sector		Mspl	Mspl
posterior cubitus	CuP	submedian or upper sector of triangle	upper sector of triangle	Cu1	Cu1
anal	A	postcosta		A	A1
first anal	A1	lower sector of triangle	lower sector of triangle	Cu2	Cu2
first anal accessory (hind wing)	A1a		subbasal sectors	A1	
anal supplement	Aspl	distal subbasal sector		bisector of anal loop	Cuspl
second anal	A2	proximal subbasal sector		A2	A2
antenodals	Ax	antecubitals	antenodals	An	Ax
postnodals	Px	postcubitals	postnodals	Pn	Px
nodus	N	nodus	nodus	n	N
subnodal crossvein	sn	subnodus	nodal cross nervure	sn	Sn
arculus	arc	arculus	arculus	arc	arc
bridge crossvein	bcv			Bq., bcv	Bx
cubito-anal crossvein	cu-a	submedian crossvein	cross nervule in lower basal cell	Ac	Cux
oblique crossvein	o			o	o
brace vein				brace vein	Bc
triangle	t	discoidal triangle	triangle	t	t

Name	Abbr.	Selys	Kirby	Needham	Tillyard 1917
subtriangle	st	internal or sub-triangle, sub-triangular space	subtriangular space	t'	st, t', ti
supratriangle	ht	hypertrigonal or supratriangular space	supratriangular space	s, ht	s, ht
discoidal field	df	posttriangular or discoidal field	posttriangular cells		d
median cell	m	basilar space	upper basal cell	m	median space
cubital cell	cu	median space	lower basal cell	cu	cubito-anal space
anal loop				anal loop	anal loop
heel of anal loop	heel			heel	
anal field (hind wing)	AF			cells behind loop	
stigma	s	pterostigma	pterostigma	st	pt
membranule	mb	membranule	membranule	mb	mb
antenodal cell	an				
quadrilateral cell	q				
subquadrilateral cell	sq				
supplementary veins	vs				

THE DRAGONFLIES OF CENTRAL AMERICA,
excl. of Mexico and the West Indies

Checklist of species

GU	Guatemala	201 species
BE	Belize	170 species
ES	El Salvador	81 species
HO	Honduras	158 species
NI	Nicaragua	54 species
CR	Costa Rica	268 species
PA	Panama	213 species

ZYGOPTERA

POLYTHORIDAE

Cora chirripa Calvert, 1907	CR
C. marina Selys, 1869	GU, BE, ES, HO, CR, PA
C. notoxantha Ris, 1918	CR, PA
C. obscura Ris, 1918	CR, PA
C. semiopaca Selys, 1878	CR, PA
C. skinneri Calvert, 1907	CR
Miocora peraltica Calvert, 1917	CR, PA

CALOPTERYGIDAE

Hetaerina americana (Fabricius, 1798)	GU, HO
H. caja (Drury, 1973)	CR, PA
H. capitalis Selys, 1873	GU, BE, ES, HO, CR, PA
H. cruentata (Rambur, 1842)	GU, BE, ES, HO, NI, CR, PA
H. fuscoguttata Selys, 1878	CR, PA
H. infecta Calvert, 1901	GU?
H. majuscula Selys, 1853	CR
H. miniata Selys, 1879	GU, HO, CR, PA
H. occisa Hagen, 1853	GU, BE, ES, HO, NI, CR, PA
H. pilula Calvert, 1901	GU, BE
H. rudis Calvert, 1901	GU
H. sempronia Hagen, 1853	BE, CR, PA
H. titia (Drury, 1773)	GU, BE, ES, HO, NI, CR
H. vulnerata Hagen, 1853	GU, HO

AMPHIPTERYGIDAE
Amphipteryx agrioides Selys, 1853 GU, HO

LESTIDAE
Archilestes grandis (Rambur, 1842) GU, BE, ES, HO, CR, PA
A. latialatus Donnelly, 1981 GU, BE, HO, NI, CR
A. neblina Garrison, 1982 CR
Lestes alacer Hagen, 1861 GU, HO, CR
L. forficula Rambur, 1842 BE, HO, CR, PA
L. henshawi Calvert, 1907 CR
L. secula May, 1993 PA
L. sigma Calvert, 1901 ES, HO, CR
L. tenuatus Rambur, 1842 GU, BE, ES, HO, CR, PA
L. tikalus Kormondy, 1959 GU, BE, HO, CR, PA

PERILESTIDAE
Perissolestes magdalenae (Williamson & Williamson, 1924) GU, BE, CR, PA
P. remotus (Williamson & Williamson, 1924) CR, PA

MEGAPODAGRIONIDAE
Heteragrion albifrons Ris, 1918 CR
H. alienum Williamson, 1919 GU, BE, HO
H. atrolineatum Donnelly, 1992 CR, PA
H. eboratum Donnelly, 1965 GU, HO
H. erythrogastrum Selys, 1886 CR, PA
H. majus Selys, 1886 CR, PA
H. mitratum Williamson, 1919 CR, PA
H. rubrifulvum Donnelly, 1992 PA
H. tricellulare Calvert, 1901 GU
H. valgum Donnelly, 1992 PA
Heteropodagrion superbum Ris, 1918 PA
Paraphlebia duodecima Calvert, 1901 GU
P. quinta Calvert, 1901 GU
Philogenia augusti Calvert, 1924 PA
P. carrillica Calvert, 1907 CR
P. championi Calvert, 1901 CR, PA
P. expansa Calvert, 1924 CR
P. lankesteri Calvert, 1924 CR
P. leonora Westfall & Cumming, 1956 PA
P. peacocki Brooks, 1989 CR
P. strigilis Donnelly, 1989 HO
P. terraba Calvert, 1907 CR
P. zeteki Westfall & Cumming, 1956 PA
Thaumatoneura inopinata McLachlan, 1897 CR, PA

PSEUDOSTIGMATIDAE

Mecistogaster linearis (Fabricius, 1777)	CR, PA
M. modesta Selys, 1860	GU, BE, HO, CR, PA
M. ornata Rambur, 1842	GU, ES, HO, NI, CR, PA
Megaloprepus caerulatus (Drury, 1782)	GU, BE, HO, NI, CR, PA
Pseudostigma aberrans Selys, 1860	GU, BE, ES, HO, CR, PA
P. accedens Selys, 1860	GU, BE, HO, CR, PA

PLATYSTICTIDAE

Palaemnema angelina Selys, 1860	GU, BE, HO
P. baltodanoi Brooks, 1989	CR
P. bilobulata Donnelly, 1992	PA
P. brucei Calvert, 1931	PA
P. chiriquita Calvert, 1931	CR
P. collaris Donnelly, 1992	CR, PA
P. cyclohamulata Donnelly, 1992	PA
P. dentata Donnelly, 1992	CR, PA
P. desiderata Selys, 1886	BE
P. distadens Calvert, 1931	CR
P. domina Calvert, 1903	GU, HO
P. gigantula Calvert, 1931	CR
P. joanetta Kennedy, 1940	PA
P. melanota Ris, 1918	CR
P. melanura Donnelly, 1992	PA
P. mutans Calvert, 1931	PA
P. nathalia Selys, 1886	GU, BE, ES, HO, CR, PA
P. paulina (Drury, 1773)	GU?, BE, HO
P. paulirica Calvert, 1931	CR
P. reventazoni Calvert, 1931	CR
P. spinulata Donnelly, 1992	PA

PROTONEURIDAE

Epipleoneura letitia Donnelly, 1992	PA
Neoneura aaroni Calvert, 1903	GU?
N. amelia Calvert, 1903	GU, BE, ES, HO, NI, CR, PA
N. esthera Williamson, 1917	CR, PA
N. paya Calvert, 1907	GU, BE
Protoneura amatoria Calvert, 1907	GU, BE, HO, CR, PA
P. aurantiaca Selys, 1886	GU, BE, CR, PA
P. cara Calvert, 1903	GU, ES, HO
P. corculum Calvert, 1907	GU, BE
P. cupida Calvert, 1903	GU, BE
P. peramans Calvert, 1902	GU, BE
P. sulfurata Donnelly, 1989	CR
Psaironeura remissa (Calvert, 1903)	GU, BE, HO, CR, PA
P. selvatica Esquivel, 1993	CR

COENAGRIONIDAE

Acanthagrion inexpectum Leonard, 1977	BE, HO, CR, PA
A. kennedii Williamson, 1916	PA
A. quadratum Selys, 1876	GU, BE, ES, HO, NI
A. speculum Garrison, 1985	CR
A. trilobatum Leonard, 1977	HO, CR, PA
Anisagrion allopterum Selys, 1876	GU, HO, NI, CR
A. kennedyi Leonard, 1937	CR, PA
A. truncatipenne Calvert, 1902	GU
Apanisagrion lais (Selys, 1876)	GU, HO
Argia adamsi Calvert, 1902	HO, CR, PA
A. anceps Garrison, 1996	GU, HO, CR
A. calida (Hagen, 1861)	GU, BE
A. chelata Calvert, 1902	BE, HO, CR, PA
A. cupraurea Calvert, 1902	HO, CR, PA
A. cuprea (Hagen, 1861)	GU, BE, HO, CR, PA
A. difficilis Selys, 1865	GU, BE, HO, CR, PA
A. eliptica Selys, 1865	BE, ES, HO, CR
A. extranea (Hagen, 1861)	GU, BE, ES, HO, CR, PA
A. fissa Selys, 1865	CR, PA
A. frequentula Calvert, 1907	GU, BE, HO, NI, CR, PA
A. funcki (Selys, 1854)	GU
A. funebris (Hagen, 1861)	ES
A. gaumeri Calvert, 1907	GU, BE, HO, CR
A. herberti Calvert, 1902	GU
A. immunda (Hagen, 1861)	BE
A. indicatrix Calvert, 1902	GU, BE, ES, HO, NI, CR, PA
A. johannella Calvert, 1907	CR
A. lacrimans (Hagen, 1861)	BE
A. medullaris Hagen, 1865	GU, CR
A. oculata Hagen, 1865	GU, BE, ES, HO, NI, CR, PA
A. oenea Hagen, 1865	GU, BE, ES, HO, NI, CR, PA
A. pallens Calvert, 1902	GU
A. pipila Calvert, 1907	GU, BE, HO
A. plana Calvert, 1902	GU
A. pocomana Calvert, 1907	GU, ES, HO, CR
A. popoluca Calvert, 1902	HO, CR, PA
A. pulla Hagen, 1865	GU, BE, ES, HO, NI, CR, PA
A. rogersi Calvert, 1902	CR, PA
A. talamanca Calvert, 1907	CR, PA
A. terira Calvert, 1907	CR
A. tezpi Calvert, 1902	GU, ES, HO, NI, CR
A. translata Hagen, 1865	GU, BE, ES, HO, NI, CR, PA
A. ulmeca Calvert, 1902	GU, BE, HO, CR, PA
A. underwoodi Calvert, 1907	CR
A. variabilis Selys, 1865	GU, CR, PA
Chrysobasis lucifer Donnelly, 1967	GU, BE, CR
Enacantha caribbea Donnelly & Alayo, 1966	GU

Enallagma civile (Hagen, 1861)	GU, HO, CR
E. novaehispaniae Calvert, 1907	GU, BE, ES, HO, NI, CR, PA
E. praevarum (Hagen, 1861)	GU, BE
E. rua Donnelly, 1968	GU, HO
Ischnura capreolus (Hagen, 1861)	GU, BE, ES, HO, NI, CR, PA
I. denticollis (Burmeister, 1839)	GU
I. hastata (Say, 1839)	GU, BE, HO, CR, PA
I. posita (Hagen, 1861)	GU, BE
I. ramburii (Selys, 1850)	GU, BE, ES, HO, NI, CR, PA
Leptobasis candelaria Alayo, 1968	GU
L. vacillans Hagen, 1877	GU, BE, ES, HO, NI, CR, PA
Metaleptobasis bovilla Calvert, 1907	GU, NI, CR
M. westfalli Cumming, 1954	CR, PA
Nehalennia minuta (Selys, 1857)	GU, BE, HO, CR, PA
Neoerythromma cultellatum (Selys, 1876)	GU, BE, ES, HO, CR, PA
Telebasis aurea May, 1992	CR, PA
T. boomsmae Garrison, 1994	BE
T. collopistes Calvert, 1902	GU, BE, HO
T. corallina (Selys, 1876)	CR
T. digiticollis Calvert, 1902	GU, BE, ES, HO, NI, CR, PA
T. filiola (Perty, 1834)	GU, BE, HO, CR, PA
T. garleppi Ris, 1918	CR
T. griffinii (Martin, 1896)	GU, BE, HO, CR, PA
T. isthmica Calvert, 1902	CR, PA
T. limoncocha Bick & Bick, 1995	PA
T. salva (Hagen, 1861)	GU, BE, ES, HO, CR, PA

ANISOPTERA

AESHNIDAE

Aeshna cornigera (Brauer, 1865)	GU, ES, HO, CR, PA
A. jalapensis Williamson, 1908	GU, ES, HO, CR, PA
A. psilus Calvert, 1947	GU, BE, HO, CR, PA
A. williamsoniana Calvert, 1905	BE, CR, PA
Anax amazili (Burmeister, 1839)	GU, BE, HO, CR, PA
A. concolor Brauer, 1865	BE, CR, PA
A. junius (Drury, 1782)	GU, BE, CR?
A. walsinghami McLachlan, 1883	GU, HO
Coryphaeschna adnexa (Hagen, 1861)	GU, BE, HO, CR, PA
C. amazonica De Marmels, 1989	CR, PA
C. apeora Paulson, 1994	BE, CR
C. diapyra Paulson, 1994	BE, CR, PA
C. viriditas Calvert, 1952	GU, BE, HO, CR, PA
Epiaeschna heros (Fabricius, 1798)	PA
Gynacantha auricularis Martin, 1909	BE, NI, CR
G. caudata Karsch, 1891	CR

G. gracilis (Burmeister, 1839)	GU, CR, PA
G. helenga Williamson & Williamson, 1930	GU, BE, HO
G. jessei Williamson, 1923	PA
G. laticeps Willimson, 1923	CR
G. membranalis Karsch, 1891	CR, PA
G. mexicana Selys, 1868	GU, BE, HO, CR, PA
G. nervosa Rambur, 1842	GU, BE, ES , HO, CR, PA
G. tibiata Karsch, 1891	GU?, CR, PA
Neuraeschna maya Belle, 1989	HO, CR
Oplonaeschna armata (Hagen, 1861)	GU, ES
Remartinia luteipennis (Burmeister, 1839)	HO, CR, PA
R. secreta (Calvert, 1952)	GU, HO
Staurophlebia reticulata (Burmeister, 1839)	GU, BE, HO, NI, CR, PA
Triacanthagyna caribbea Williamson, 1923	GU, BE, HO, CR, PA
T. dentata (Geijskes, 1943)	PA
T. ditzleri Williamson, 1923	GU, CR, PA
T. satyrus (Martin, 1909)	BE, CR, PA
T. septima (Selys, 1857)	GU, BE, ES, HO, CR, PA

GOMPHIDAE

Agriogomphus tumens (Calvert, 1905)	GU, BE, CR, PA
Aphylla angustifolia Garrison, 1986	GU, BE
A. obscura (Kirby, 1899)	CR, PA
A. protracta (Hagen, 1859)	GU, BE, NI, CR
Archaeogomphus furcatus Williamson, 1923	CR
Desmogomphus paucinervis (Selys, 1873)	CR, PA
Epigomphus armatus Ris, 1918	CR
E. camelus Calvert, 1905	CR
E. clavatus Belle, 1980	GU
E. compactus Belle, 1994	PA
E. corniculatus Belle, 1989	CR
E. echeverrii Brooks, 1989	CR
E. houghtoni Brooks, 1989	CR
E. jannyae Belle, 1993	PA
E. maya Donnelly, 1989	BE
E. quadracies Calvert, 1903	GU, CR, PA
E. subobtusus Selys, 1878	GU, ES, HO, CR
E. subquadrices Kennedy, 1947	PA
E. subsimilis Calvert, 1920	CR
E. tumefactus Calvert, 1903	CR
E. verticicornis Calvert, 1908	CR
E. westfalli Donnelly, 1986	NI
Erpetogomphus bothrops Garrison, 1994	GU, ES, NI, CR
E. constrictor Ris, 1917	ES, HO, NI, CR
E. elaphe Garrison, 1994	GU, HO, CR
E. elaps Selys, 1857	GU
E. eutainia Calvert, 1905	GU, BE, ES, HO, CR
E. leptophis Garrison, 1994	BE

E. ophibolus Calvert, 1905   GU, BE

Species	Distribution
E. ophibolus Calvert, 1905	GU, BE
E. sabaleticus Williamson, 1918	PA
E. schausi Calvert, 1919	GU, CR
E. tristani Calvert, 1912	CR, PA
Perigomphus pallidistylus (Belle, 1972)	CR, PA
Phyllocycla breviphylla Belle, 1975	GU, BE, NI
P. elongata (Selys, 1857)	GU
P. speculatrix Belle, 1975	GU, BE
P. volsella (Calvert, 1905)	GU, BE, CR, PA
Phyllogomphoides appendiculatus (Kirby, 1899)	CR, PA
P. bifasciatus (Hagen, 1878)	GU, ES, HO, NI, CR
P. burgosi Brooks, 1989	CR
P. duodentatus Donnelly, 1979	GU, BE, ES, HO
P. insignatus Donnelly, 1979	PA
P. litoralis Belle, 1984	PA
P. pugnifer Donnelly, 1979	GU, BE, CR
P. suasillus Donnelly, 1979	GU
P. suasus (Selys, 1859)	GU, BE, NI, CR
Progomphus anomalus Belle, 1973	CR, PA
P. clendoni Calvert, 1905	GU, BE, ES, HO, NI, CR, PA
P. longistigma Ris, 1918	CR
P. mexicanus Belle, 1973	BE, CR
P. pygmaeus Selys, 1873	GU, CR
P. risi Williamson, 1920	GU
P. zonatus Hagen, 1854	GU, BE

CORDULEGASTRIDAE

Species	Distribution
Cordulegaster godmani McLachlan, 1978	GU, CR

CORDULIIDAE

Species	Distribution
Neocordulia batesi (Selys, 1871)	CR, PA
N. campana May & Knopf, 1988	CR, PA
N. griphus May, 1992	CR

LIBELLULIDAE

Species	Distribution
Anatya guttata (Erichson, 1848)	GU, BE, HO, CR, PA
Brachymesia furcata (Hagen, 1861)	GU, BE, ES, HO, CR, PA
B. herbida (Gundlach, 1888)	BE, ES, HO, CR, PA
Brechmorhoga nubecula (Rambur, 1842)	BE, CR, PA
B. pertinax (Hagen, 1861)	GU, HO, CR, PA
B. praecox (Hagen, 1861)	GU, BE, ES, HO, NI, CR, PA
B. rapax Calvert, 1898	GU, BE, ES, HO, CR, PA
B. tepeaca Calvert, 1907	BE
B. vivax Calvert, 1906	GU, BE, ES, HO, CR, PA
Cannaphila insularis Kirby, 1889	GU, BE, ES, HO, NI, CR, PA
C. mortoni Donnelly, 1992	CR, PA

C. vibex (Hagen, 1861)	GU, BE, ES, HO, CR, PA
Dythemis maya Calvert, 1906	GU, ES
D. multipunctata Kirby, 1894	GU, BE, ES, HO, CR, PA
D. sterilis Hagen, 1861	GU, BE, ES, HO, NI, CR, PA
D. velox Hagen, 1861	GU
Elasmothemis cannacrioides (Calvert, 1906)	GU, BE, HO, CR, PA
Elga leptostyla Ris, 1911	PA
Erythemis attala (Selys, 1857)	GU, BE, HO, CR, PA
E. carmelita Williamson, 1923	PA
E. credula (Hagen, 1861)	BE, HO, PA
E. haematogastra (Burmeister, 1839)	BE, HO, NI, CR, PA
E. mithroides (Brauer, 1900)	GU, CR, PA
E. peruviana (Rambur, 1842)	GU , HO, NI, CR, PA
E. plebeja (Burmeister, 1839)	GU, BE, ES, HO, CR, PA
E. simplicicollis (Say, 1839)	BE, ES, HO, NI, CR
E. vesiculosa (Fabricius, 1775)	GU, BE, ES, HO, NI, CR, PA
Erythrodiplax abjecta (Rambur, 1842)	CR
E. andagoya Borror, 1942	CR
E. berenice (Drury, 1773)	GU, BE, CR, PA
E. castanea (Burmeister, 1839)	BE, CR
E. connata (Burmeister, 1839)	CR, PA
E. famula (Erichson, 1848)	CR
E. fervida (Erichson, 1848)	GU, BE, ES, HO, NI, CR, PA
E. funerea (Hagen, 1861)	GU, BE, ES, HO, NI, CR, PA
E. fusca (Rambur, 1842)	GU, BE, ES, HO, CR, PA
E. kimminsi Borror, 1942	CR, PA
E. lativittata Borror, 1942	PA
E. umbrata (Linnaeus, 1758)	GU, BE, ES, HO, NI, CR, PA
E. unimaculata (de Geer, 1773)	PA
Idiataphe amazonica (Kirby, 1889)	BE, HO, CR, PA
I. cubensis (Scudder, 1866)	GU, BE, HO, CR
Libellula croceipennis Selys, 1868	GU, BE, ES, HO, CR
L. foliata (Kirby, 1889)	GU, HO, CR, PA
L. gaigei Gloyd, 1938	GU, BE
L. herculea Karsch, 1889	GU, BE, ES, HO, CR, PA
L. mariae Garrison, 1992	CR
Macrodiplax balteata (Hagen, 1861)	BE
Macrothemis aurimaculata Donnelly, 1984	GU, CR
M. delia Ris, 1913	GU, CR
M. extensa Ris, 1913	CR
M. fallax May, 1998	BE, PA
M. hemichlora (Burmeister, 1839)	GU, BE, ES, HO, CR, PA
M. imitans Karsch, 1890	GU, BE, HO, CR, PA
M. inacuta Calvert, 1898	GU, BE, HO, NI, CR, PA
M. inequiunguis Calvert, 1895	GU, BE, ES, HO, CR, PA
M. musiva Calvert, 1898	BE, HO, CR, PA
M. nobilis Racenis, 1957	PA
M. pseudimitans Calvert, 1898	GU, BE, ES, HO, CR, PA
Miathyria marcella (Selys, 1857)	GU, BE, ES, HO, NI, CR, PA

M. simplex (Rambur, 1842)	GU, BE, HO, CR, PA
Micrathyria aequalis (Hagen, 1861)	GU, BE, ES, HO, NI, CR, PA
M. atra (Martin, 1897)	GU, BE, HO, CR, PA
M. caerulistyla Donnelly, 1992	PA
M. catenata Calvert, 1909	CR
M. debilis (Hagen, 1861)	GU, BE, ES, HO
M. dictynna Ris, 1919	GU, BE, CR, PA
M. didyma (Selys, 1857)	GU, BE, HO, CR, PA
M. dissocians Calvert, 1906	GU, BE
M. hagenii Kirby, 1890	BE HO, CR, PA
M. laevigata Calvert, 1909	GU, CR, PA
M. mengeri Ris, 1919	GU, BE, CR, PA
M. ocellata Martin, 1897	GU, BE, HO, NI, CR, PA
M. pseudeximia Westfall, 1992	GU, HO, CR, PA
M. schumanni Calvert, 1906	GU, ES, CR, PA
M. tibialis Kirby, 1897	CR, PA
Nephepeltia chalconota Ris, 1919	GU, BE, HO, CR
N. leonardina Racenis, 1953	PA
N. phryne (Perty, 1834)	GU, BE, HO, CR, PA
Oligoclada heliophila Borror, 1931	PA
O. umbricola Borror, 1931	GU, CR
Orthemis aequilibris Calvert, 1909	PA
O. biolleyi Calvert, 1906	GU, BE, CR, PA
O. cultriformis Calvert, 1899	CR, PA
O. discolor (Burmeister, 1839)	CR
O. ferruginea (Fabricius, 1775)	GU, BE, ES, HO, NI, CR, PA
O. flavopicta Kirby, 1889	PA
O. levis Calvert, 1906	GU, BE, ES, HO, NI, CR, PA
Pachydiplax longipennis (Burmeister, 1839)	BE
Paltothemis lineatipes Karsch, 1890	GU, BE, ES, HO, NI, CR
Pantala flavescens (Fabricius, 1798)	GU, BE, ES, HO, NI, CR, PA
P. hymenaea (Say, 1839)	GU, BE, HO, CR, PA
Perithemis domitia (Drury, 1773)	GU, BE, ES, HO, CR, PA
P. electra Ris, 1930	GU, HO, CR, PA
P. mooma Kirby, 1889	GU, BE, ES, HO, NI, CR, PA
Planiplax phoenicura Ris, 1912	PA
P. sanguiniventris Calvert, 1907	GU, BE, ES
Pseudoleon superbus (Hagen, 1861)	GU, ES, HO, NI, CR
Rhodopygia cardinalis (Erichson, 1848)	PA
R. hinei Calvert, 1907	GU, BE, CR, PA
Sympetrum corruptum (Hagen, 1861)	BE, HO
S. illotum (Hagen, 1861)	GU, ES, HO, CR, PA
S. nigrocreatum Calvert, 1920	CR
Tauriphila argo (Hagen, 1869)	GU, BE, HO, CR, PA
T. australis (Hagen, 1867)	GU, BE, ES, HO, CR, PA
T. azteca Calvert, 1906	GU, CR
Tholymis citrina Hagen, 1867	GU, BE, HO, CR, PA
Tramea abdominalis (Rambur, 1842)	GU, BE, HO, CR, PA
T. binotata (Rambur, 1842)	GU, BE, HO, CR, PA

T. calverti (Muttkowski, 1910)	GU, BE, ES, HO, CR, PA
T. insularis Hagen, 1861	CR
T. onusta (Hagen, 1861)	GU, BE, HO, CR, PA
Uracis fastigiata (Burmeister, 1839)	GU, HO, NI, CR, PA
U. imbuta (Burmeister, 1839)	GU, BE, ES, HO, NI, CR, PA
U. turrialba Ris, 1919	BE, CR
Zenithoptera americana (Linnaeus, 1758)	CR

Total number of species: 378

References: PAULSON (1982, 1997), RAMIREZ et.al. (1999), WESTFALL & MAY (1996)

Key to the families of Odonata

Plate IV, p. 21

1	Front- and hindwing of different shape, base of hindwing broader than that of forewing	**2**
1'	Front- and hindwing similar in shape, narrowed at base	**6**
2	Triangles of the front and hind wings dissimilar in shape, that of the forewing with the long axis at right angles to the length of the wing, that of the hind wing with its long axis parallel to the length of the wing (Fig. F1), and much nearer to the arculus than that of the forewing; eyes meeting on top of the head	**5**
2'	Triangles of front and hind wings usually of similar shape and subequally distant from the arculus in each, except in some Gomphidae (Fig. F2)	**3**
3	Eyes widely separated at top of head, abdominal segments without lateral carinae	**Gomphidae** (p. 74)
3'	Eyes meeting at top of head	**4**
4	Eyes meeting only at one point, body black with yellow markings, wings without a crossvein under the proximal end of the pterostigma	**Cordulegastridae** (p. 89)
4'	Eyes meeting in a line (longer than 1 mm); wings with a crossvein under the proximal end of the pterostigma (Fig. F3)	**Aeshnidae** (p. 66)
5	Anal loop in hind wing foot-shaped (Fig. F1)	**Libellulidae** (p. 92)
5'	Anal loop in hind wing rounded or sac-shaped (Fig. F2)	**Corduliidae** (p. 90)
6	Antenodal crossveins 7 or more	**7**
6'	Antenodal crossveins 2 (up to 5 in *Thaumatoneura*)	**9**
7	7-10 antenodal crossveins	**Amphipterygidae** (p. 28)
7'	20 or more antenodal crossveins	**8**
8	Median vein arises from the upper end of the arculus; anterior side of triangle concave (Fig. F5)	**Polythoridae** (p. 22)
8'	Median veins R4+5 and MA arise from the lower end or the middle of the arculus; anterior side of the triangle convex (Fig. F4)	**Calopterygidae** (p. 24)
9	R4+5 arises closer to arculus than to nodus (Fig. F7)	**Lestidae** (p. 30)
9'	R4+5 arises closer to nodus than to arculus (Fig. F6)	**10**
10	Pterostigma absent or with crossveins	**Pseudostigmatidae** (p. 39)
10'	Pterostigma present and consisting of one single cell (uncrossed)	**11**
11	Quadrilateral reaching the hind margin of wing (Fig. F9); abdomen very long and slender	**Perilestidae** (p. 33)
11'	Quadrilateral not reaching the hind margin of wing (Fig. F10); abdomen not as above	**12**
12	Anterior and posterior sides of quadrilateral subequal in length; A1 absent or very short (Figs. F10, F11)	**13**
12'	Anterior side of quadrilateral distinctly shorter than posterior side; A1 well developed, reaching the level of the nodus or beyond (Figs. F6, F12)	**14**
13	Distal end of anal vein meets the hind margin of the quadrilateral; cubito-anal crossvein present (Fig. F 10)	**Platystictidae** (p. 41)
13'	Anal vein absent, or reduced to a small arc at the hind margin of wing; cubito-anal crossvein absent (Fig. F11)	**Protoneuridae** (p. 46)

14   Distance between base of wing and nodus less than one-third the length of the
        wing (Fig. F12)..**Megapodagrionidae** (p. 34)

14'   Distance between base of wing and nodus about one-third the length of the wing
        (Fig. F6)..**Coenagrionidae** (p. 49)

After: ESQUIVEL (1991)

Plate IV

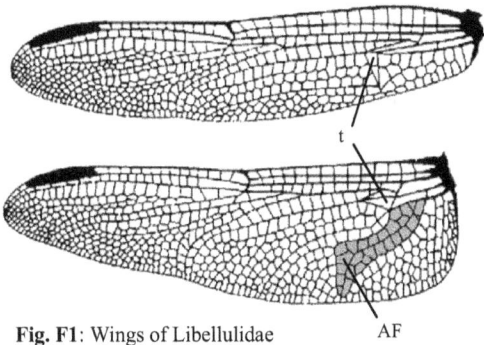

Fig. F1: Wings of Libellulidae
(*Orthemis ferruginea*)

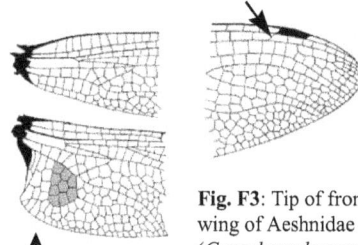

Fig. F2: Wing bases of Aeshnidae
(*Coryphaeschna* spec.)

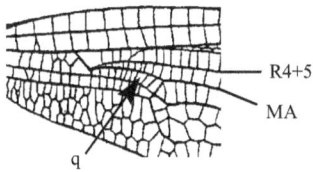

Fig. F3: Tip of front
wing of Aeshnidae
(*Coryphaeschna* spec.)

Fig. F4: Base of frontwing of Calopterygidae
(*Hetaerina* spec.)

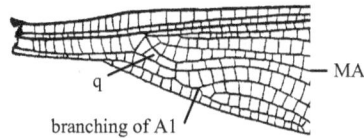

Fig. F5: Base of front wing of Polythoridae
(*Cora semiopaca*)

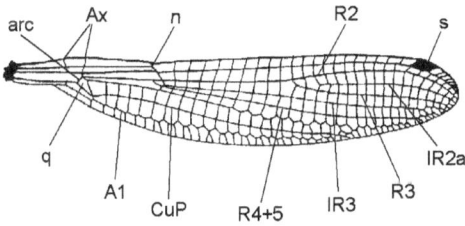

Fig. F6: Frontwing of Coenagrionidae (*Argia* spec.)

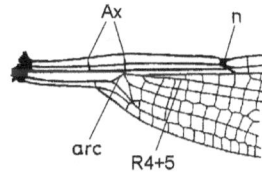

Fig. F7: Base of frontwing of Lestidae
(*Archilestes grandis*)

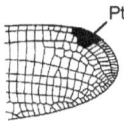

Fig. F8: Tip of frontwing
of Pseudostigmatidae
(*Mecistogaster modesta*)

Fig. F9: Base of front wing
of Perilestidae
(*Perissolestes remotus*)

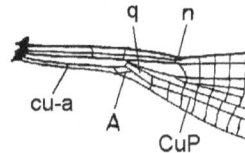

Fig. F10: Base of front wing of
Platystictidae (*Palaemnema* spec.)

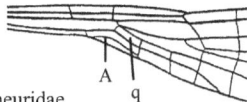

Fig. F11: Base of
frontwing of Protoneuridae
(*Neoneura amelia*)

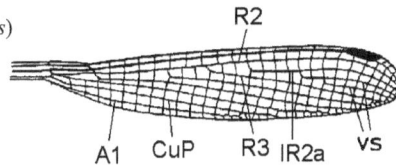

Fig. F12: Front wing of Megapodagrionidae
(*Heteragrion erythrogastrum*)

Polythoridae

Medium sized damselflies of rather robust appearance, which sometimes have black, white, or yellow wing markings. They live at small shaded streams in primary forests and are generally found by logs lying over the stream, where the females lay eggs in the wet wood (DUNKLE, in litt.). They are generally not very abundant at their reproductive sites.

Key to the genera of Polythoridae

1   A1 branched (Fig. F5); more than one row of cells between CuP and A1 and between A1 and the wing margin; quadrilateral of hind wing scarcely longer than that of fore wing ....***Cora*** (p. 22)

1'   A1 unbranched; one row of cells between CuP and A1 and between A1 and the wing margin; quadrilateral with 2-4 cross-veins, usually three in fore wing and four in hind; many supplementary sectors; dark spot near apex of hind wing, extending to wing margin ....***Miocora*** (p. 23)

After: MUNZ (1919)

Cora

GONZÁLEZ SORIANO & VERDUGO GARZA (1984a, b) studied the reproductive behavior of *C. marina*. Males usually perform precopulatory courtship flights. Females oviposit within the territory of the male, which guards them from its perching site (,,long distance guarding"). Often there is only one male present at a specific reproductive site, even at large sites their number seldomly exceeds six individuals.

Males (see Figs. Co1-6, p.23, for thorax patterns)

1   Abdominal segment 2 dorsally almost all pale ....*marina*

1'   Abdominal segment 2 dorsally at least half dark **2**

2   With a dark band on each wing extending from just beyond the nodus to or almost to the apex ....*semiopaca*

2'   Wings entirely hyaline or with a small brown or black apical mark **3**

3   Mesepimeron mostly pale (yellow or blue), sometimes with a black streak **4**

3'   Mesepimeron mostly dark (brown or black) **6**

4   Dorsal carina with a narrow black stripe (Fig. Co2); pale areas on mesepisternum and mesepimeron either yellow-orange or blue-gray

       ....***notoxantha***

4'   Dorsal carina with a broad black stripe which may or may not bulge (Fig. Co4); pale areas on mesepisternum and mesepimeron blue-gray **5**

5   Wings entirely hyaline; fore wing 26-31 mm ....*chirripa chirripa*

5'   Each wing with a dark brown or black mark from proximal end of pterostigma to wing apex; fore wing 31-35 mm ....***chirripa donnellyi***

6   Fore wing long, 30-34 mm; metepisternum without 2 elongate pale stripes but
        with posterior half pale; pterothorax sternum with paired black spots_____***skinneri***

6'   Fore wing much shorter, 22-26 mm; metepisternum with 2 elongate pale stripes
        separated by black or brown_____***obscura***

After: BICK & BICK (1990)

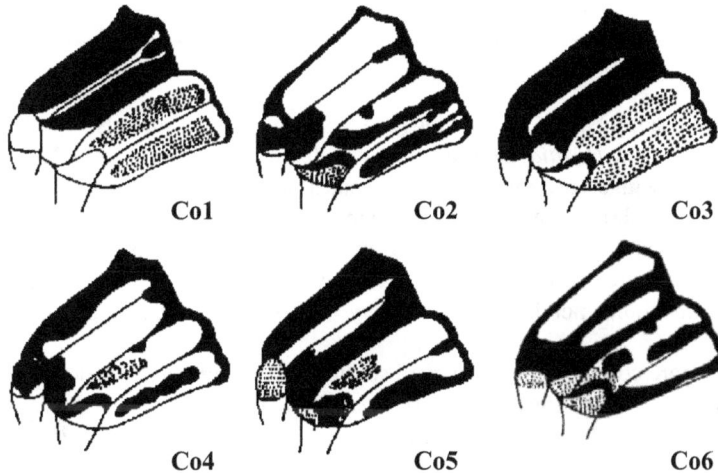

Figs. Co1 - Co6: *Cora* sp., left lateral view of male pterothorax
Co1 - *Cora semiopaca*, **Co2** - *C. notoxantha*, **Co3** - *C. obscura*, **Co4** - *C. c. chirripa*, **Co5** - *C. skinneri*,
Co6 - *C. marina*

Miocora

Miocora peraltica

This is the only species of the genus in Central America. Eyes are dark brown,
somewhat bluish below. Thoracic dorsum blackish; humeral suture, most of
metepisternum and much of metepimeron pale bluish green, but metepisternum and
metepimeron each with a blackish stripe. Abdomen black; a spot on each side of
segment one, a longitudinal stripe on each side of segment two, and a small spot on
each side of segment three are pale green (CALVERT, 1917; KENNEDY, 1940).

Calopterygidae

Plate V (p. 25) & VI (p. 29)

In Central America, this family of medium sized damselflies is represented only by the genus *Hetaerina*. These damselflies are characterized by a rather dense wing venation, having numerous antenodal crossveins and many intercalated veins, which alternate with the primary longitudinal veins. The pterostigma is poorly developed, usually not more than a single small cell.

Hetaerina

The wings of the males usually have a conspicuous red basal spot, sometimes also a red or brown spot at their tips. In *H. titia* there are many different color forms in males, some of them with hind wings entirely blackish. Females show a light brown wing coloration. The bodies of both sexes are bronzy or coppery colored. Males often perch with the body slightly inclined above the horizontal at streams and rivers of different sizes, preferably in forested areas. Most species are closely associated with their larval habitat and are generally found not far from it at any time in their life (see studies mentioned below).

The behavior of some more common species is well studied, e.g. *H. americana* (BICK & SULZBACH, 1966; JOHNSON, 1961, 1962, 1963), *H. titia* (JOHNSON, 1961, 1963), or *H. macropus* (EBERHARD, 1986). Males hold territories near suitable oviposition sites, often performing remarkable display flights (not in all species). Females oviposit underwater without the male which usually waits above the surface.

Males

1	Inferior appendage small, consisting of a basal plate and a median process; distal process reduced to a small tubercle or point 0.25 or less the length of superior appendage **2**
1'	Inferior appendage with a fully developed distal process at least 0.25 as long as superior appendage, its tip armed with 1-3 small teeth **5**
2	Basal 0.33 of superior appendage with a strong, almost transverse elevated ridge; mesal margin of appendage almost straight and undifferentiated (Fig. Ca1) ....*miniata*
2'	No strongly elevated transverse ridge across basal 0.33 of superior appendage; mesal margin of superior appendage at least with a mesal lobe (except in *H. titia*) **3**
3	Superior appendage with widened median lobe at basal 0.5 of appendage, this lobe entire; basal 0.5 of appendage wider than distal 0.5 (Fig. Ca2).......*capitalis*
3'	Superior appendage with widened median lobe at distal 0.5 of appendage (Figs. Ca3 & 4), this lobe weakly notched (*H. majuscula*) or strongly bilobed (*H. infecta*); basal 0.5 of appendage in mediodorsal view narrower than distal 0.5...**4**
4	Median lobe weakly notched, its posterior margin not well demarcated from distal fossa (Fig. Ca3), latter longer than wide in mediodorsal view; thorax

Plate V

Ca1	**Ca2**	**Ca3**	**Ca4**	**Ca5**	**Ca6**	**Ca7**
H. miniata	*H. capitalis*	*H. majuscula*	*H. infecta*	*H. rudis*	*H. sempronia*	*H. occisa*

Ca8	**Ca9**	**Ca10**	**Ca11**	**Ca12**	**Ca13**	**Ca14**
H. pilula	*H. titia*	*H. americana*	*H. cruentata*	*H. vulnerata*	*H. fuscoguttata*	*H. caja*

Ca15	**Ca16**	**Ca17**	**Ca18**	**Ca19**	**Ca20**	**Ca21**
H. rudis	*H. titia*	*H. caja*	*H. cruentata*	*H. capitalis*	*H. cruentata*	*H. miniata*

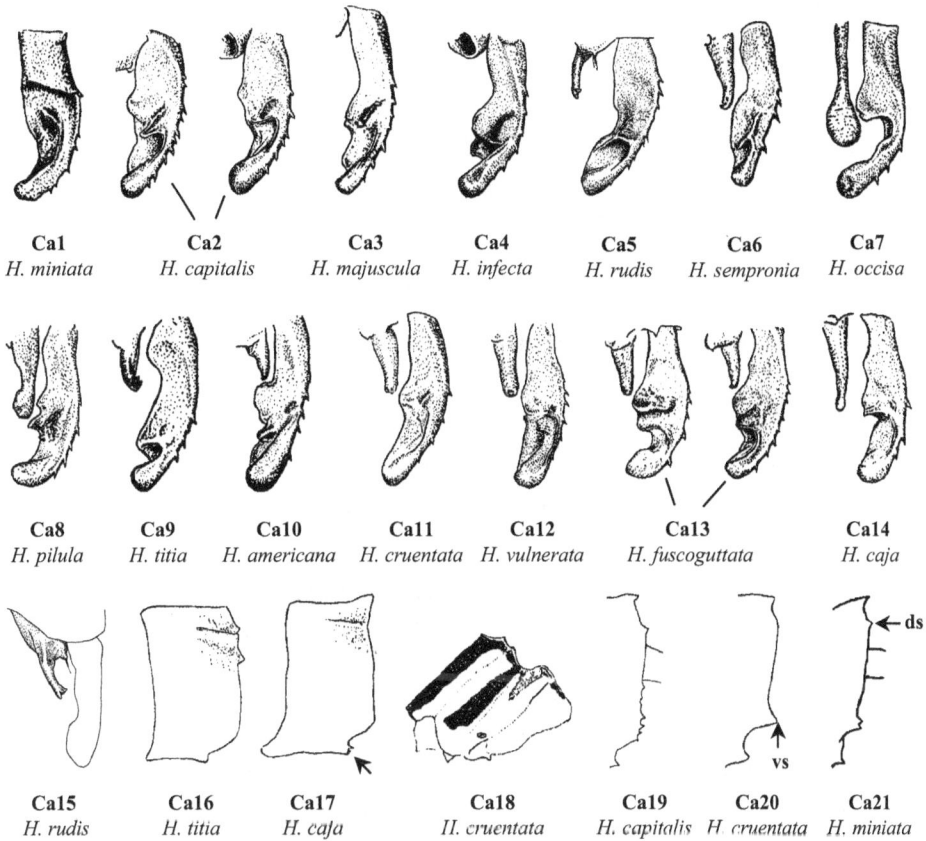

Figs. Ca1-Ca18: *Hetaerina* sp., male characters;
Ca1-14: Dorsal view of right caudal appendages,
Ca15: Lateral view of caudal appendages;
Ca16 & 17: Dorsolateral view of abdominal segment 10;
Ca18: Thorax, left lateral view.

Figs. Ca19-22: *Hetaerina* sp.,
posterior margin of abd. segment 10
of females, left lateral view;
vs - ventrolateral spine, **ds** - dorso-
lateral spine, **ms** - middorsal spine

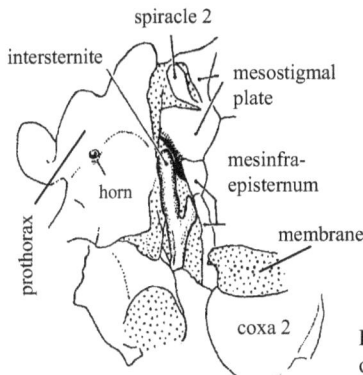

Ca22
H. sempronia

Fig. Ca24: *H. occisa*, prothorax
and anterior part of synthorax of female, dorsal view.

Fig. Ca23: *H. occisa*, prothorax and anterior part of synthorax
of female, left lateral view.

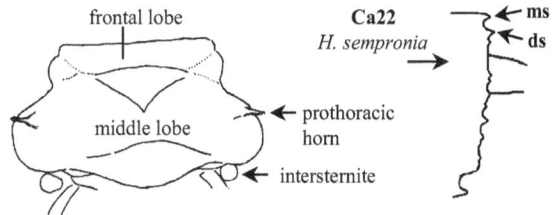

cupreous black with metallic reflections; abdominal segments 2-7 often with slight metallic purple luster..***majuscula***

4'   Posterior margin of median lobe usually well demarcated from distal fossa (Fig. Ca4), structure wider than long in mediodorsal view; thorax dark cupreous often with metallic reflections on dorsal 0.33 of thoracic stripes; abdominal segments 2-7 with dark brown luster, never with purple luster...............................***infecta***

5   Labrum dark metallic green; wings with pterostigma..**6**

5'   Labrum pale, dark brown, black or dark medially with white lateral spots; lacking any metallic luster; wings with or without pterostigma...............................**7**

6   Distal process of inferior appendage in lateral view distinctly bifurcate; large species (hind wing 35 mm); all wings with apical red spot; appendages as in Fig. Ca5...***rudis***

6'   Distal process of inferior appendages armed with 2-3 small recurved teeth; in lateral view, this structure appearing blunt; smaller species (hind wing up to 32 mm); fore wing tips hyaline, hind wing tips with brown spot; appendages as in Fig. Ca6...***sempronia***

7   Distal process of inferior appendage enlarged distally (Figs. Ca7 & 8)................**8**

7'   Distal process of inferior appendage of various shapes, but not with tip enlarged **9**

8   Distal process of inferior appendage 0.66 the length of superior appendage; tip of distal process dorsoventrally flattened, forming a racquet shape; median lobe of superior appendage weakly bilobate (Fig. Ca7)...***occisa***

8'   Distal process of inferior appendage about 0.50 to 0.60 the length of superior appendage; tip of distal process clublike; median lobe strongly bidentate (Fig. Ca8)..***pilula***

9   Tip of abdominal segment 10 armed with a well-developed triangular tooth just lateral to middorsal carina (Fig. Ca16); appendages as in Fig. Ca9...............***titia***

9'   Tip of abdominal segment 10 lacking accessory tooth lateral to middorsal carina (Fig. Ca17)...**10**

10   Anterior margin of median lobe meeting medial margin of superior appendage at approximately a 90 degree angle (Fig. Ca10); with a pterostigma........***americana***

10'   Anterior margin of median lobe, if present, forming an obtuse angle with medial margin of superior appendage; with or without a pterostigma...............................**11**

11   Thorax with some metallic green on metepisternum...**12**

11'   Thorax predominantly metallic red to dull brown interspersed with pale lines along thoracic sutures, no metallic green on metepisternum...............................**13**

12   Metallic green on thorax confined to a well-defined inverted wedge-shaped spot on dorsal end of metepisternum (Fig. Ca18); superior appendage with mesal margin gently curved with a small median lobe; in mediodorsal view, width of basal 0.33 of appendage about as wide as distal 0.33 (Fig. Ca11)........***cruentata***

12'   Metallic green present on metepisternum (though it may be confined to dorsalmost portion of border; darker stripe extending full length of metepisternum), mesepimeron and mesepisternum; superior appendage (Fig. Ca12) with mesal margin variously shaped, not gently curved as in *H. cruentata*; median lobe, if present, variously shaped, but not as in Fig. Ca11........***vulnerata***

13   Middorsal carina on abdominal segment 10 forming a sharp spine surpassing posterior margin of segment, abdominal appendages as in Fig. Ca13 ....***fuscoguttata***

13'   Middorsal carina on abdominal segment 10 ending posteriorly as an elevated
        ridge, or disappearing altogether **14**

14   Distal process of inferior appendage terminating in a single recurved tooth; distal
        process usually 0.5 or more as long as superior appendage (Fig. Ca14) ....***caja***

14'   Distal process of inferior appendage with a blunt, truncate tip, terminating in 1-2
        small recurved teeth; distal process usually 0.33 as long as superior appendage
        (Fig. Ca12) ....***vulnerata***

Females

1   Anterior arm of intersternite froming a dorsally directed fingerlike digit (Fig.
        Ca23); prothorax with laterally extended horns (Fig. Ca24); thorax as in Fig.
        Ca38 ....***occisa***

1'   Dorsal end of intersternite planar, flattened, not digitlike; prothorax lacking
        laterally directed horns (except in some *H. vulnerata*) **2**

2   Abdominal segments with a definite color pattern as follows: each segment
        dorsally dark metallic green or cupreous (obscure brown in old individuals), with
        a thin pale middorsal line; lateral margins of dorsal green straight, contrasting
        with pale sides; dorsal green often sending an offshoot connecting with ventral
        margin at posterior 0.2 of segment (these marks often isolated on anterior 1-3
        segments)(Fig. Ca25) ....***americana***

2'   Abdominal segment 1 usually with a discernible pattern, but succeeding
        segments obscure, often totally dark and lacking any pattern **3**

3   With a pterostigma **4**

3'   Without a pterostigma **12**

4   Dark thoracic coloration dark metallic red **5**

4'   Dark thoracic coloration metallic green **6**

5   Entire labrum and postclypeus dark metallic blue green; epicranium often with
        metallic blue green luster; tip of abdominal segment 10 armed dorsally with
        three spines, one middorsal, one each side dorsolateral, each as long as dorsal
        (Fig. Ca22); intersternite as in Fig. Ca32 ....***sempronia***

5'   Labrum ivory with an inverted black triangular mark along basal margin;
        postclypeus black with slight metallic red luster, never with metallic blue green;
        epicranium black with slight metallic red luster; tip of abdominal segment 10
        armed dorsally with three spines, but middorsal much the largest (Fig. Ca21);
        intersternite as in Fig. Ca29 ....***miniata***

6   Tip of abdominal segment 10 armed with three small spines, one middorsal and
        one each side dorsolateral (e.g. Fig. Ca19) **7**

6'   Tip of abdominal segment 10 armed with only one long middorsal spine or none,
        no dorsolateral spines (Fig. Ca20) **11**

7   Intersternite linear, with a small anterior shoulder followed by a longer, bluntly
        pointed or slightly rounded posterior branch (Fig. Ca26); thorax as in Fig. Ca34
        ....***capitalis***

7'   Intersternite broad, flat, appearing slightly tridentate (Fig. Ca30), slightly
        bidentate or rounded (Fig. Ca27, Ca31) or with anterior arm longer than
        posterior arm (Fig. Ca30) **8**

8   Intersternite with anterior branch longer than posterior branch, anterior branch
        acute, sometimes with a small, isolated sclerite dorsal to intersternite (Fig.
        Ca30); smaller species (hind wing to 32 mm); thorax as in Fig. Ca39 ....***pilula***

8' Intersternite rounded dorsally, slightly bilobate or tridentate, never with an extended anterior branch, larger species (hind wing 34 mm or more) **9**

9 Intersternite appearing slightly tridentate, dorsal margin prominent, not emarginate (Fig. Ca28); thorax as in Fig. Ca36 .... *majuscula*

9' Intersternite appearing smoothly rounded or slightly emarginate dorsally **10**

10 A small, isolated sclerite dorsal to intersternite; intersternite widened dorsally, its surface appearing concave (Fig. Ca27) .... *infecta*

10' Without an isolated sclerite dorsal to intersternite; intersternite not especially widened dorsally, its surface not strongly concave (Fig. Ca31) .... *rudis*

11 Apical margin of abdominal segment 10 with no spine (Fig.Ca20); with a small sclerite dorsal to intersternite; side of pronotum in some individuals with a laterally directed pointed process; wings hyaline with some yellow-orange coloration basally .... *vulnerata*

11' Apical margin of abdominal segment 10 with a spine (Fig. Ca16); no isolated sclerite dorsal to intersternite; side of pronotum without a laterally directed process; wings in some individuals hyaline or nearly so, but more often completely dark, almost black .... *titia*

12 Dark thoracic coloration metallic red .... *fuscoguttata*

12' Dark thoracic coloration metallic green or bronze **13**

13 Ventrolateral margin of abdominal segment 10 armed posteriorly with a broadly pointed spine (Fig. Ca20) **14**

13' Ventrolateral margin of abdominal segment 10 armed posteriorly with 2-3 or more small pointed spines (Fig. Ca17) .... *caja*

14 Metallic green mesepisternal stripe connecting with humeral suture (Fig. Ca41); with a round scleritized structure just dorsal to intersternite; dorsal area of intersternite wider than in *H. cruentata*, its anterior arm deflected .... *vulnerata*

14' Metallic green mesepisternal stripe not connecting with humeral suture (Fig. Ca35); no round scleritized structure just dorsal to intersternite, or, if present, then this structure small, barely noticeable; dorsal area of intersternite narrower than in *H. vulnerata*, its anterior arm never deflected .... *cruentata*

After: GARRISON (1990)

Amphipterygidae

Amphipteryx agrioides

These are large sized, robust damselflies with an abdomen length of about 55 mm. The thorax of the male is green with black stripes, the abdomen black with yellow spots. The female is very similar to the male but somewhat smaller. Species of this genus can be found at seepages and small streams in rain or cloud forest habitats (GONZÁLEZ SORIANO, 1991).

Plate VI

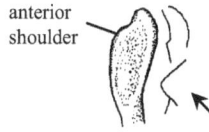

anterior shoulder

Fig. Ca25: *Hetaerina americana*, female abdomen, left lateral view

Ca26: *H. capitalis*

Ca27	**Ca28**	**Ca29**	**Ca30**	**Ca31**	**Ca32**
H. infecta	*H. majuscula*	*H. miniata*	*H. pilula*	*H. rudis*	*H. sempronia*

Figs. Ca26-32: *Hetaerina*, female intersternite, left lateral view

Ca33: *H. caja*   **Ca34**: *H. capitalis*   **Ca35**: *H. cruentata*   **Ca36**: *H. majuscula*   **Ca37**: *H. miniata*

Ca38: *H. occisa*   **Ca39**: *H. pilula*   **Ca40**: *H. rudis*   **Ca41**: *H. vulnerata*

Figs. Ca33-41: *Hetaerina*, thorax of female, left lateral view

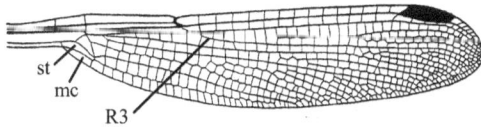

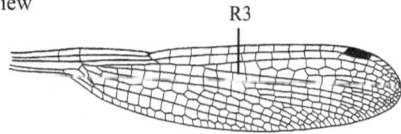

R3

st
mc
R3

Fig. Le1: *Archilestes grandis* (Lestidae), forewing   **Fig. Le2**: *Lestes sp.* (Lestidae), forewing

Le3: *L. alacer*   **Le4**: *L. forficula*   **Le5**: *L. henshawi*   **Le6**: *L. secula*   **Le7**: *L. sigma*   **Le8**: *L. tikalus*

Fig. LE3-8: Male abdominal appendages of *Lestes* spp., dorsal view

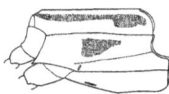 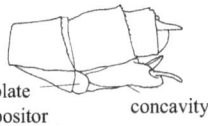

basal plate
of ovipositor   concavity

Fig. Pe1: *Perissolestes remotus*, male superior appenadage, dorsal view

Fig. Le9: *L. tikalus,* thorax color pattern

Fig. Le10: *L. secula*, apex of female abdomen

Fig. Pe2: *P. magdalenae*, male superior appendage, dorsal view

Lestidae

Plate VI, p. 29

Medium sized damselflies (40-60 mm abdomen length). Males are of bluish and metallic green coloration, females are brownish. As an exception in the Zygoptera, they hold their wings half outspread while perching (also found in the Megapodagrionidae).

Key to the genera of Lestidae

1	Inner end of the quadrilateral in the fore wing almost one-half the length of the posterior side; R2+3 forking less than two cells beyond the nodus; third antenodal space almost twice the length of the first (Fig. Le1)

Archilestes (p. 30)

1'	Inner end of the quadrilateral in the fore wing one-third or less the length of the posterior side; R2+3 forking more than two cells beyond the nodus; third antenodal space not more than one and one-half times the length of the first (Fig. Le2)

Lestes (p. 31)

After: MUNZ (1919)

Archilestes

The species of this small, mostly Neotropical genus are distinguished from *Lestes* by their larger size and robust stature, and by the rather short, broad quadrangle. They are most often found on streams in forested areas, generally in pools or backwaters with slow current (GARRISON, 1982; WESTFALL & MAY, 1996). RAMÍREZ (1994) provided more detailed ecological notes on the three species of the genus, representing the most comprehensive information available so far. According to him, *A. grandis* is a quite common species, at least in Costa Rica, occuring at a variety of habitats ranging from pools to streams, and can be found even in very polluted environments. Both *A. latialatus* and *A. neblina* preferably inhabit streams in forested areas. The former occurs at altitudes usually not higher than 1000 m ASL, whereas the latter has been recorded between 900 and 1500 m ASL, and seems to occur only in the Braulio Carrillo National Park and the Monteverde Cloud Forest Reserve, Costa Rica.

Males

1	Marginal cell posterior to subtriangle (mc in Fig. Le1) in hindwing enlarged at border so that anal border of wing is strongly convex ....*latialatus*
1'	Marginal cell posterior to subtriangle in hindwing not greatly enlarged at border, anal border of wing gently convex **2**
2	Posterior half of superior appendages greatly expanded, truncate and leaf-like in appearance; large species, abdomen excluding appendages 55 mm, hind wing 42 mm ....*neblina*

2'   Posterior half of superior appendages linear, similar in width to basal half; smaller species, abdomen excluding appendages less than 50 mm, hind wing less than 40 mm; inferior abdominal appendages divergent, not parallel sided
....*grandis*

After: GARRISON (1982)

Lestes

Medium or large sized damselflies. Adults rest with the body inclined well below the horizontal, and the wings half outspread. Wings are usually hyaline, strongly petiolated, and with a long, narrow pterostigma. Abdomen typically entirely dark dorsally, with rear of head, thorax, and basal as well as terminal abdominal segments tending to become pruinose with age. Species of the genus are usually restricted to lentic waters. They prefer small lakes and ponds with marshy margins and abundant emergent vegetation (WESTFALL & MAY, 1996). Reproductive behavior of the Central American species is almost completely unknown.

Males

1   Inferior appendages sigmoid, their tips diverging (Fig. Le7).....................*sigma*
1'   Inferior appendages not sigmoid, their tips parallel or converging...................**2**
2   Superior appendages, in dorsal view, scythe-like, their medial margins without large teeth (Fig. Le6) ....*secula*
2'   Superior appds. not so, their medial margins with at least one prominent tooth
....**3**
3   Inferior appds. barely more than ½ length of superior appds., or less...............**4**
3'   Inferior appds. distinctly more than ½ (typically more than 2/3) length of superior appds..**5**
4   Femora and superior surface of head mostly yellow to yellow brown (head darkens with age); mesepisterna each with metallic green stripe, not contiguous with mid-dorsal carina (mid-dorsal area may become dark brown)........*tenuatus*
4'   Femora and posterior surface of head mostly black; mesepisterna each with uniformly black stripe, contiguous witrh mid-dorsal carina; abdominal appendages as in Fig. Le5...*henshawi*
5   Superior appds., in dorsal view, with distinct subapical, medial tooth in addition to similar basal tooth (Fig. Le8); metallic green mesepisternal stripe abruptly widened posteriorly (Fig. Le9)...*tikalus*
5'   Superior appds., in dorsal view, with basal, medial tooth only, usually followed by more or less serrated margin; mesepisternal stripe uniform in width or only slightly and gradually widened posteriorly...**6**
6   Mesepisterna each with narrow, metallic green stripe, not contiguous with mid-dorsal carina; inferior appds. nearly as long as superiors, extending well beyond posterior limit of medial dilation of latter (Fig. Le4).......................*forficula*
6'   Mesepisterna each with broad black stripe, contiguous with mid-dorsal carina; inferior appds. not more than ¾ as long as superiors, extending little, if at all, beyond posterior limit of medial dilation of latter..**7**

7   Superior appds., in dorsal view, with medial serrated dilation a well developed lobe terminated posteriorly by a distinct notch; inferior appds. not distinctly curved inward (Fig. Le3)..*alacer*

7'   Superior appds. in dorsal view, with medial serrated dilation less prominent, not terminating posteriorly in a distinct notch; inferior appds. curved or slanted inward..*simplex*

Females

1   Basal plate of ovipositor rounded at posteroventral margin; ventral margin of valvula 3 of ovipositor often with distinct concavity (Fig. Le10)...................**2**

1'   Basal plate of ovipositor with posteroventral corner acutely angulate, often with an acute tooth; ventral margin of valvula 3 of ovipositor nearly straight or convex throughout..**4**

2   Posterior surface of head yellow or yellow brown; thoracic venter without dark streak along metapleural carina; length of abdominal segment 3 about 6 times its height at midlength..*tenuatus*

2'   Posterior surface of head black; thoracic venter with dark streak just ventromedial to metapleural carina, in addition to dark anterior and posterior spots adjacent to carina; length of abdominal segment 3 about 5 times its height at midlength..**3**

3   Ventral margin of valvula 3 of ovipositor with distinct concavity; pale stripes of mesothorax often not sharply demarked and dark mesepimeral stripes either much shorter than mesepimeron or largely brwonish and obscure; pronotum not black; hindwing shorter than 25 mm..*secula*

3'   Ventral margin of valvula 3 of ovipositor nearly straight; pale stripes of mesothorax sharply demarked and dark mesepimeral stripes black and extending almost full length of mesepimeron; pronotum largely black; hindwing usually longer than 25 mm..*henshawi*

4   Mesepisterna with distinctly metallic green stripes or spots.........................**5**

4'   Mesepisterna without metallic green areas, dark areas black, sometimes with bronze reflections..**6**

5   Mesepisternal stripe usually wider than 0.3 mm at narrowest point, abruptly expanded posteriorly; metallic green mesepimeral stripe, if present, wedge-shaped, widest anteriorly..*tikalus*

5'   Mesepisternal stripe usually narrower than 0.3 mm at narrowest point, only slightly and gradually widened posteriorly; metallic green mesepimeral stripe, if present, linear..*forficula*

6   Mesepisterna and mesepimera with dark spots or streaks extending much less than length of each sclerite or, in old specimens, entire mesothorax becoming dark and heavily pruinose..*sigma*

6'   Mesepisterna and mesepimera with complete, dark stripes extending nearly full length of respective sclerites, this pattern never entirely obscured.............*alacer*

After: MAY (1993)

Perilestidae

Plate VI, p. 29

Both species occuring in Central America belong to the genus *Perissolestes*. They are characterized by a very long and slender abdomen. Sides of thorax with black and green stripes. Abdomen dark with light greenish spots near basal border of segments. Rather rare damselflies, which are restricted to rain forests. The adults are usually found in deep shade, perching on branches of small trees near small streams (GONZALEZ SORIANO & DEL PILLAR VILLEDA, 1978; WILLIAMSON & WILLIAMSON, 1924).

Perissolestes

Males

1   Superior appendage with the ento-basal lobe nearer the base of the appendage than the apex of the middle lobe (Fig. Pe1)⟋⟋⟋⟋⟋***remotus***

1'   Superior appendage with the ento-basal lobe beyond midlevel of appendage, nearer apex of middle internal lobe than of base of appendage; appendages very hairy to beyond the middle (Fig. Pe2)⟋⟋⟋***magdalenae***

Females

1   Hind lobe of prothorax bearing an erect median spine; pterostigma black; vulvar styles attached at the level of the cerci⟋⟋⟋***magdalenae***

1'   Prothorax not spined; point of attachment of valvular styles distal to the level of the apices of the cerci ⟋⟋⟋***remotus***

After: KENNEDY (1941)

34

Megapodagrionidae

Plate VII, p. 35

Medium sized damselflies. Males of many species are conspicuously colored, often with red or yellow abdomens. Females are brownish. Adults perch with wings half outspread (as an exception in Zygoptera, also found in the Lestidae) and abdomen inclined below the horizontal. In contrast to the Lestidae, they preferably occur along streams and small rivers in primary forests (ESQUIVEL, 1991).

Key to the genera of Megapodagrionidae

1	Antenodals 3-5, postnodals over 50; venation very complex, many supplementary sectors from R2 to the posterior margin of the wing; about 10 crossveins under the pterostigma *Thaumatoneura* (p. 38)
1'	Antenodals 2-3, postnodals generally not over 30; crossveins under the pterostigma 6 or fewer **2**
2	IR3 arising at or very near the subnodus; R4+5 at least one cell before the subnodus; 5 crossveins under the pterostigma (Fig. Me1) *Philogenia* (p. 37)
2'	IR3 arising at least one cell beyond subnodus; R4+5 usually not before the subnodus, at least not a cell's length before it **3**
3	Distance between CuP and A1 at the wing margin as great as that from the distal edge of the quadrilateral to the end of A1; quadrilateral very long, almost or quite reaching the subnodus (Fig. Me2) *Paraphlebia* (p. 36)
3'	CuP and A1 usually ending near together at the wing margin (Fig. Me3) **4**
4	Supplementary sectors rather numerous, some between IR3 and R4+5 *Heteropodagrion* (p. 34)
4'	No supplementary sectors between IR3 and R4+5 (Fig. Me3) *Heteragrion* (p. 34)

After: MUNZ (1919)

Heteragrion

These are variously colored damselflies which live „...along shaded streams from which they rarely wander for any distance" (WILLIAMSON, 1919b). Accordingly, larvae of the species studied so far (e.g. RAMÍREZ, 1992) can be found in primary and secondary dry or wet forest streams at sites with little or no current, where organic matter accumulates. Females are dull brown and are easily overlooked when perching low over ground on the forest floor. Like other shade seeking odonates of tropical forests, the behavior of adults of *Heteragrion* is characterized by a very low flight activity (see SHELLY, 1982).
The reproductive behavior of *H. alienum* was studied by GONZÁLEZ SORIANO & VERDUGO GARZA (1982). Males hold territories at the stream bank, not including the oviposition sites. After copulation the pair oviposits in tandem position preferably on floating leaves.

Plate VII

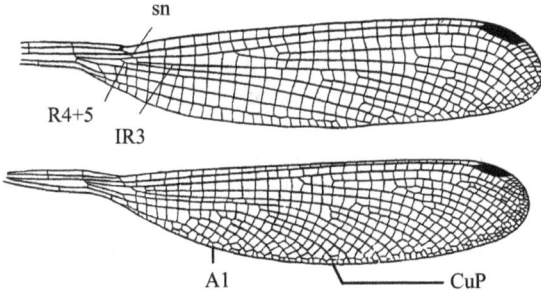

Fig. Me1: *Philogenia terraba,* fore wing

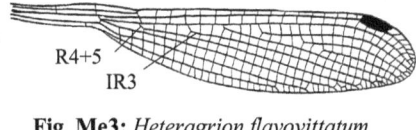

Fig. Me2: *Paraphlebia duodecima,* hind wing

Fig. Me3: *Heteragrion flavovittatum,* hind wing

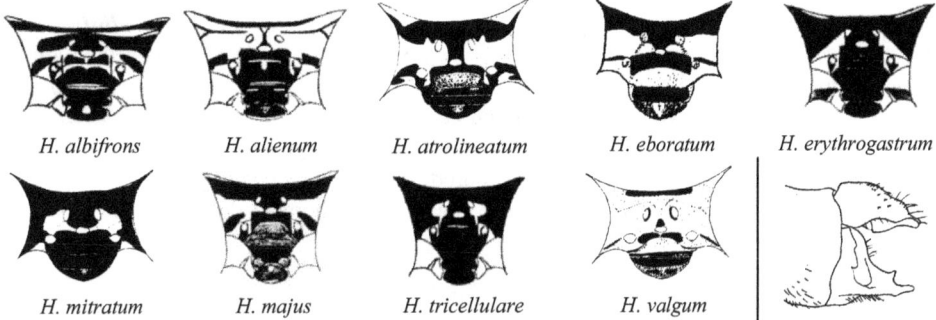

H. albifrons   *H. alienum*   *H. atrolineatum*   *H. eboratum*   *H. erythrogastrum*

H. mitratum   *H. majus*   *H. tricellulare*   *H. valgum*

Me5: *P. augusti*

Fig. Me4: Females of *Heteragrion*, color pattern of head in dorsal view

Me6: *P. carrillica*   **Me7:** *P. championi*   **Me8:** *P. leonora*   **Me9:** *P. peacocki*   **Me10:** *P. strigilis*

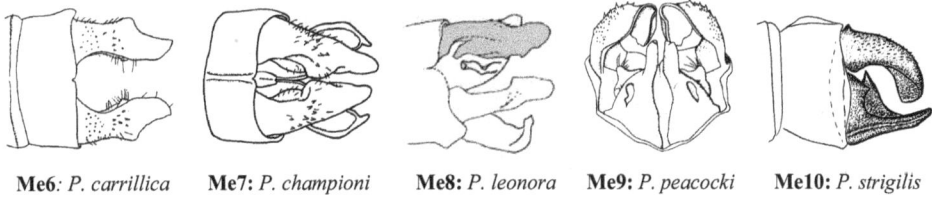

Fig. Me5-10: Males of *Philogenia*, apex of abdomen
Me5 - dorsolateral view; **Me6 & 8** - lateral view; **Me7** - dorsal view; **Me9 & 10** - ventral view

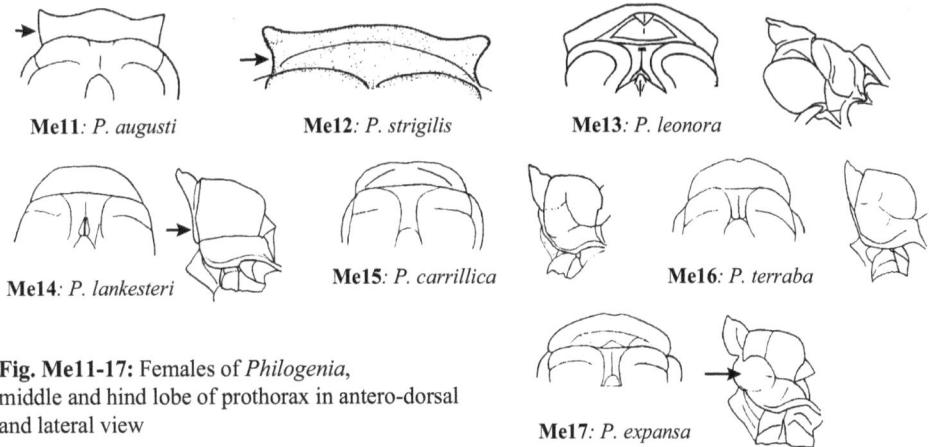

Me11: *P. augusti*   **Me12**: *P. strigilis*   **Me13**: *P. leonora*

Me14: *P. lankesteri*   **Me15**: *P. carrillica*   **Me16**: *P. terraba*

Fig. Me11-17: Females of *Philogenia*,
middle and hind lobe of prothorax in antero-dorsal
and lateral view

Me17: *P. expansa*

Males (for females see **Fig. Me4**)

1   Abdomen above predominantly pale, no segment as much as one-fourth
       definitely black; segments 1 and 2 without any definite black areas; segments 8-
       10 largely or entirely pale or obscure **2**
1'   Abdominal patterns various but segments 1 and 2 with definite dark areas or at
       least some segments definitely largely or entirely black above **5**
2   Abdomen largely red **3**
2'   Abdomen reddish yellow or flesh to almost red, segments 3-6 with distinct apical
       black areas; face yellow **4**
3   Face black; abdomen without black; thorax dominantly black ..... *erythrogastrum*
3'   Face bright light yellow; abdomen with some segments narrowly apically black
       .... *albifrons*
3''   Face red; thorax pale brown .... *rubrifulvum*
4   Abdomen 47-51.5 mm long; usually three postquadrangular cells ..... *tricellulare*
4'   Smaller; usually two postquadrangular cells .... *alienum*
5   Abdominal segments 3-6 dorsally and laterally largely black, 8-10 and the
       appendages to near the apex pale coloured **6**
5'   Abdomen basally pale, perhaps with black rings; thoracic dorsum pale or pale
       striped with black **7**
6   Entire face shining white .... *eboratum*
6'   Face golden yellow .... *majus*
7   Terminal abdominal segments dark on dorsum .... *mitratum atroterminatum*
7'   Terminal abdominal segments pale on dorsum **8**
8   Thorax vividly striped with yellow and black; dorsum of the head pale brown
       with dark spots .... *valgum*
8'   Thorax pale without vivid stripes; dorsum of head orange (front) with a velvet
       black stripe (rear) .... *atrolineatum*

Derived from: DONNELLY (1965b, 1992), WILLIAMSON (1919b)

Heteropodagrion

Heteropodagrion superbum

Medium sized damselfly. Males with thorax dark brown, legs incl. spines bright red,
very slender, bright red abdomen, abdominal appendages reddish. Wings yellowish,
pterostigma bright red (RIS, 1918).

Paraphlebia

1   Thorax ventrally black; males with the apical fifth of the wings dark brown,
       beginning well in front of the pterostigma, the inner edge straight; superior
       appendages bent inward at two-fifths their length .... *quinta*

1'   Thorax ventrally yellow; males with the apical 12th to 14th of the forewings (less on the hind) dark brown, beginning at the inner end or at the middle of the pterostigma, the inner edge somewhat convex, superior appendages uniformly curved throughout, not angulate..*duodecima*

After: CALVERT (1908)

Philogenia

„The following pattern of coloration is characteristic of the genus: Dorsal surface of the head blackish... Rear of the head pale yellowish, a blackish spot opposite the attachment of each maxilla. Labrum, external surface of each mandible and adjacent part of gena pale green, greenish-blue or yellow, narrowly marginated with black. Thorax with ... most of the mesinfraepisternum ... black; paler parts of the thorax dull brown or yellow; pectus yellow. Abdomen blackish; a spot ... or longitudinal stripe on each side of segment one and two, a transverse basal ring, often interrupted mid-dorsally, on three to seven; a lateral stripe extending caudad from this ring on three and often on the following segments, pale green or yellowish; one or more of the hindmost segments pruinose dorsally in older males." (CALVERT, 1924)
Most species are nearly identical except for the male terminal appendages. They often inhabit dark primary forests along small streams.

Males (*expansa* & *lankesteri* only known from females)

1   In dorso-mesal view, the meso-ventral process of superior appendage with an extension at its proximal end directed towards base of appendage (Fig. Me8)....**2**
1'   Without such an extension...**6**
2   In ventral view, distal halves of inferior appendages convergent; in lateral view, inferior with a small subapical tooth...*terraba*
2'   Without the above combination of characters...**3**
3   In lateral view, inferior appendages bent 90° upward just anterior to apex (Fig. Me8)...*leonora*
3'   In lateral view, inferior appendages not bent sharply upward, at most curving gradually upward apically...**4**
4   In lateral view, inferior appendages taper to apex, with a small dorsal spine in its distal half (Fig. Me5)...*augusti*
4'   In lateral view, inferiors truncate at apex, without a distinct dorsal spine in its distal half..**5**
5   Superior appendages in dorsal view with a prominence at mid-length of lateral margin (Fig. Me6), in lateral view with their meso-ventral margin produced ventrad as a distinct process in the distal three-fifths of the appendage...*carrillica*
5'   Superior appendages in dorsal view with a prominence at mid-length of the mesal margin (Fig. Me7), in lateral view with their meso-ventral margin not produced ventrad as a distinct process, but forming a slightly convex curve
      ....*championi*

6   Inferior appendages as long or slightly longer than superiors, divaricating in
        ventral view, and with a prominent subbasal, dorsally directed spine when
        viewed laterally (Fig. Me10)..*strigilis*
6'   Inferior appendages distinctly shorter than superior appendages.....................**7**
7   In ventral view, inferior appendages strongly divergent apically.................*zeteki*
7'   Inferiors not strongly divergent, with a ventral cavity (Fig. Me9).........*peacocki*

Females (*championi*, *peacocki* and *zeteki* only known from males)

1   Hind lobe of pronotum bisinuate in antero-dorsal view, with lateral angles
        directed dorsad (Figs. Me11 & 12)...**2**
1'   Hindlobe of pronotum as a whole convex in antero-dorsal view (Figs. Me13-17),
        not produced in an acute angulate process on each side..............................**3**
2   In antero-dorsal view, sides of hind lobe of pronotum more convexly shaped
        (Fig. Me11)..*augusti*
2'   In antero-dorsal view, sides of hind lobe of pronotum more concavely shaped
        (Fig. Me12)..*strigilis*
3   Sides of hind lobe of pronotum almost straight and vertical in antero-dorsal view
        (Fig. Me13)..*leonora*
3'   Sides of hind lobe of pronotum convex and rounded in antero-dorsal view (Figs.
        14-17)..**4**
4   Hind margin of hind pronotal lobe not continued on the epimeron; apex of wings
        not dark brown..**5**
4'   Hind margin of hind pronotal lobe continued on the epimeron (Fig. Me14)
        ....*lankesteri*
5   Hind margin of proepimeron in lateral view not concealed by any convex
        production of the sclerite itself (Figs. Me15 & 16)...........*carrillica & terraba*
5'   Hind margin of proepimeron in lateral view concealed by a strongly convex
        production of the sclerite (Fig. Me17)..*expansa*

Derived from: BICK & BICK (1988), BROOKS (1989), CALVERT (1924), DONNELLY
(1989a), MAY (1989), WESTFALL & CUMMING (1956)

Thaumatoneura

Thaumatoneura inopinata

This large sized damselfly (abdomen 55 mm long) is easily recognized by a broad black
band across each wing in the male, although there are also males which lack the wing
bands. The females have clear wings except for a black apical spot. They are generally
found around waterfalls, where the larvae live among roots in small water basins
(CALVERT & CALVERT, 1917). The species is restricted to Costa Rica and Panama.

Pseudostigmatidae

These damselflies are easily identified by their very long and slender abdomen of about 80-120 mm in length. The wings often have white, yellow, or dark blue tips. Their flight is very slow, and they are able to catch small web-building spiders or their wrapped prey from the webs without getting entangled in them. Adults can be found around light gaps in primary forests. The larvae live in small water-filled treeholes or tank bromeliads. Some species, especially *Megaloprepus coerulatus*, can be locally abundant.
FINCKE (1984, 1992a, 1992b) did some extensive studies on the reproductive behavior of pseudostigmatid species on Barro Colorado Island (Panama).

Key to the genera of Pseudostigmatidae

1   Wings broad, inferior sector of triangle and the short sector curved and much branched; superior appendages of the male shorter than the inferiors; quadrilateral free; pterostigma square or rectangular..........***Megaloprepus*** (p. 39)

1'   Wings narrow, inferior sector of triangle and the short sector almost straight, unbranched, or the latter slightly branched; superior appendages of male longer than the inferiors_____**2**

2   Postcostal area of two cell rows, at least in its middle portion
        --***Pseudostigma*** (p. 40)

2'   Postcostal area of a single row of cells_____***Mecistogaster*** (p. 39)

After: CALVERT (1908)

Megaloprepus

Megaloprepus coerulatus

This is the only species of the genus, which is distributed widely from Central to South America. Adults of *M. coerulatus* have a wing span of up to 180 mm and an abdomen as long as 100 mm. Both sexes are recognized by a wide, dark blue band across the distal third of the wings. During the reproductive period (mainly in wet and early dry season) males occupy territories around breeding sites.

Mecistogaster

1   No false pterostigma ('pseudostigma'); tips of all wings with an opaque spot whose upper surface varies from pale yellow through orange-yellow to dark olive according to age, older males with ventral sides of the wing tips completely black_____*ornata*

1'   Pseudostigma present_____**2**

2   No terminal opaque spot on the wings; pterostigma occupying two rows of cells
        on the forewings and one row on the hind wings......................................***modesta***

2'   Wing tips milky white (young), older individuals with black pseudostigma;
        mature males with clear wingtips (wings with brownish tinge); mature females
        with wing tips faint opaque white..***linearis***

Compiled from: CALVERT (1908), FINCKE (1984)

Pseudostigma

1   Opaque spot on the wings (false pterostigma) reaching to the extreme tip; in the
        males it stops short of reaching the end of the median vein by a distance very
        much less than its own length; in the females it reaches IR_{2a} or R3; superior
        appendages of male bent strongly downward in their apical half, inferior
        appendages moderately developed..***abberans***

1'   Opaque spot on the wings not reaching to the extreme tip, in the male it stops
        short of reaching the end of the median vein by a distance greater than its own
        length, in the females it reaches to the principal sector or to one row of cells
        below it; superior appendages of male not bent downward in their apical half;
        inferiors rudimentary..***accedens***

After: CALVERT (1908)

Platystictidae

Plate VIII, p. 43

Medium sized damselflies. Body generally black with light blue spots on labrum, thorax and abdomen. Some have black wing tips. They can be found near rivers and streams in primary forests. They mostly perch quietly close to the ground and are therefore easily overlooked. Sometimes they rest with wings half outspread (similar to Lestidae and Megapodagrionidae). In the New World there is only one described genus, *Palaemnema*.

Palaemnema

The reproductive behavior of these damselflies is exeptional. As reported by GONZÁLEZ ET.AL. (1982) and GONZÁLEZ & VERDUGO (1984) for *P. desiderata*, the reproductive activity starts just around sunrise, and two hours later all individulas return to inactivity for the rest of the day, perching motionless close to the ground. During the period of mating, males occupy territories in and around bushes at the stream bank. After a short copulation, females oviposit within the territory of the male, which often succeeds in mating with additional females in succession. Among the males, there also exist alternative mating strategies, such as trying to get females without the effort of establishing and defending territories, which is similar to mating systems found in calopterygid species (e.g. SIVA-JOTHY & TSUBAKI, 1989).

Panama

Males

1	Propleura pale	2
1'	Propleura dark (Fig. PL1)	5
2	Antehumeral stripe absent	3
2'	Antehumeral stripe present	*nathalia*
3	Tip of superior appendage excised; internal sub-apical spine on inferior appendage (Fig. PL14)	*spinulata*
3'	Tip of superior appendage smooth; no spine as above (Fig. PL5)	4
4	Hind lobe of pronotum blue; pterostigma widened apically	*collaris*
4'	Hind lobe of pronotum dark; pterostigma not widened apically	*cyclohamulata*
5	Antehumeral stripe absent	6
5'	Antehumeral stripe or spot present	7
6	Prominent dorsal spine on superior appendage (Fig. PL12); pterostigma as broad as long, red	*mutans*
6'	Dorsal spine on superior appendage reduced to small step (Fig. PL9); pterostigma longer than broad, dark	*joanetta*
7	Antehumeral stripe a short, tapered spot	8
7'	Antehumeral stripe the length of mesepisternum	9

8   Dorsum of abdominal segment 8 blue in its posterior ¾; very low dorsal spine on
        the superior appendage; dorsum of pterothorax with metallic red reflections
        ....*bilobulata*

8'   Dorsum of abdominal segment 8 chiefly black; distinct, acute dorsal spine on the
        superior appendage (Fig. PL4); thoracic dorsum bronze violet.................*brucei*

9   Tip of abdomen blue; triangular ventral projection on basal part of widened tip
        of superior appendage (Fig. PL6)..*dentata*

9'   Tip of abdomen black; no triangular ventral projection as described above
        ....*melanura*

Females

(Note: pale colours can often be blue or yellow, depending on age of individual;
ultraviolet light may help to clarify colour patterns)

1   Propleura pale...2
1'   Propleura dark..5
2   Antehumeral stripe absent...3
2'   Antehumeral stripe present..4
3   Conspicuous pale blue on hind lobe of pronotum; pterostigma widened apically;
        ovipositor not surpassing superior appendage..*collaris*
3'   Dark hind lobe of pronotum; pterostigma not widened apically; ovipositor
        surpassing superior appendage...*spinulata*
4   Dorsum of 8-10 blue; ovipositor surpasses superior appendage....*cyclohamulata*
4'   Dorsum of 8-10 dark except for a pale spot on 9; ovipositor does not surpass
        superior appendage...*nathalia*
5   Antehumeral stripe absent or a spot...6
5'   Antehumeral stripe present...*melanura*
6   Antehumeral spot present...7
6'   Antehumeral spot absent..8
7   Ovipositor surpasses superior appendage...*bilobulata*
7'   Ovipositor does not surpass superior appendage..*dentata*
8   Abdominal segment 9 dark..*mutans*
8'   Abdominal segment 9 pale..*joanetta*
8''   Abdominal segment 9 dark, with a pair of blue or yellow spots which may or
        may not be confluent on the mid-dorsal line..*brucei*

Costa Rica and farther north

Males

1   Propleuron chiefly black or dark brown (Fig. PL1)..2
1'   Propleuron pale blue or yellowish; apex of wings uncoloured or fainty smoky
       13

2   Superior appendages with no sinus or notch between the superior tooth and the
        widening of the ventral margin when viewed from above (Fig. PL8); inferiors
        only 0.6 as long as the superiors; pale (blue) antehumeral stripe extending the

propleuron

antehumeral stripe

Plate VIII

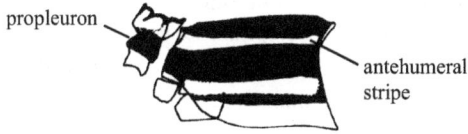

Fig. PL1: *Palaemnema dentata*, lateral view of pro- and pterothorax

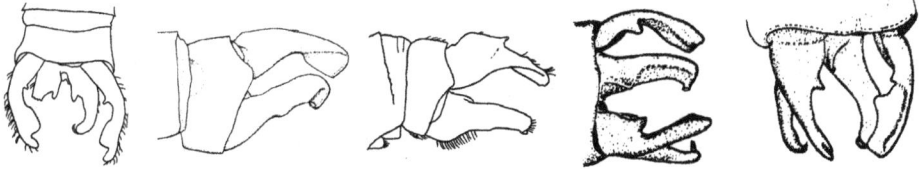

PL2: *P. angelina*   **PL3:** *P. baltodanoi*   **PL4:** *P. brucei*   **PL5:** *P. cyclohamulata*   **PL6:** *P. dentata*

PL7: *P. distadens*   **PL8:** *P. domina*   **PL9:** *P. johanetta*   **PL10:** *P. paulina*   **PL11:** *P. paulirica*

PL12: *P. mutans*   **PL13:** *P. reventazoni*   **PL14:** *P. spinulata*   **PL15**

Figs. PL2-14: Abdominal appendages of *Palaemnema* males, dorsal or dorso-lateral view

Fig. PL15: *P. paulina*, lateral view of thorax

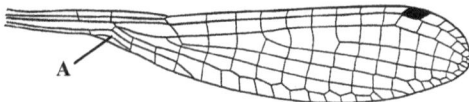

A

Fig. PR1: *Neoneura amelia*, hind wing

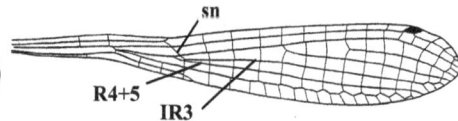

sn

R4+5

IR3

Fig. PR2: *Protoneura capillaris*, hind wing

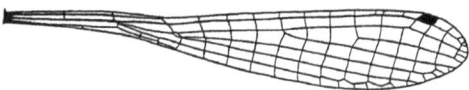

Fig. PR3: *Psaironeura remissa*, hind wing

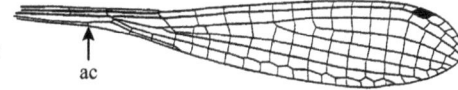

ac

Fig. PR4: *Epipleoneura letitia*, fore wing

PR5: *Protoneura cara*   **PR6:** *P. cupida*   **PR7:** *Psaironeura remissa*   **PR8:** *P. selvatica*

Figs. PR5-8: Protoneuridae, male anal appendages, lateral view

entire length of the mesepisternum, narrower than the combined fused humeral and first lateral dark stripes; a dark stripe on the metepimeron ventral to and fusing with that on the second lateral suture ....*domina*

2' No such combination of characters **3**

3 Abdominal segments 8 and 9 chiefly blue on dorsum **4**

3' Abdominal segment 8 chiefly black; no dark metepimeral or metasternal stripes **11**

4 With a conspicuous triangular ventral projection on basal part of widened tip of superior appendage (Fig. PL6) ....*dentata*

4' No such triangular projection **5**

5 Metepimeron with a brown or black longitudinal stripe dorsal to the latero-ventral carina, extending in most cases (except in Costa Rican forms) across the carina and onto the metasternum (Fig. PL15) ; mature males with dark brown at the apices of the wings (except in *P. distadens*) **6**

5' Metepimeron with no brown or black longitudinal stripe, no dark stripes or markings on the metasternum behind the third legs; mature males with no dark brown at apices of wings **10**

6 Apex of superior appendages not deeply excised at 90°, the superior margin not prolonged as an overhanging process **7**

6' Apex of superior appendages deeply excised at 90°, superior margin prolonged as an overhanging process (Fig. PL2); blue antehumeral stripe extending for almost the entire length of the mesepisternum; superior appendage with superior tooth at 0.3-0.38 of appendage length; distal one-eighth to one-fifteenth of wings dark brown in mature males, uncoloured in immatures ....*angelina*

7 Metasternum with no dark stripes or markings behind 3rd legs; blue antehumeral stripe only a little wider in its lower half than at midheight **8**

7' Metasternum with dark stripes or markings behind the 3rd legs **9**

8 Superior appendages with the superior tooth at 0.33-0.43 of the appendage length, apex of appendage slightly excised (Fig. PL11); blue antehumeral stripe at midheight wider than the black humeral, except in those males in which dark humeral and first lateral stripes are fused; distal one-twelfth to one-sixth of wings dark brown in mature males ....*paulirica*

8' Superior appendages with the superior tooth at 0.53-0.62 of the appendage length, apex of appendage rounded, not at all excised (Fig. PL7); blue antehumeral stripe at midheight narrower than the black humeral; no males yet found with dark brown at wing apices ....*distadens*

9 Blue antehumeral stripe markedly widened in its lower half as compared with width at midheight; superior tooth of superior appendages at 0.3 of appendage length (Fig. PL10) ....*paulina*

9' Blue antehumeral stripe only a little wider in its lower half as compared with width at midheight ....*desiderata*

10 Abdomen 1.6 times (or less) as long as hind wing; a pale antehumeral stripe or spot present on at least the lower end of the mesepisternum; mesinfraepisternum chiefly black, pale inferiorly ....*mutans*

10' Abdomen 1.8 times as long as hind wing; no pale antehumeral stripe or spot; mesinfraepisternum pale blue; superior appendages curved strongly from base to apex in profile view ....*gigantula*

11 Dorsum of abdominal segment 9 blue; superior appendages with apex excised at 90° (Fig. PL12); blue antehumeral stripe very short ....*mutans*

11'   Dorsum of 9 black **12**

12   Antehumeral stripe present; abdominal appendages as in Fig. PL3 ...._**baltodanoi**_

12'   Antehumeral stripe absent; abdominal appendages as in Fig. PL13 ..._**reventazoni**_

13   Pale antehumeral stripe or line present ...._**nathalia**_

13'   Pale antehumeral stripe or line absent, mesinfraepisternum mostly blue **14**

14   Abdominal segments 8 to 9 or 10 blue on dorsum (except for anterior one-seventh or less of eight, which is black) **15**

14'   Abdominal segments 8 to 10 blackish ...._**melanota**_

15   Hind lobe of prothorax pale blue; superior appendage with apex not excised
       ...._**collaris**_

15'   Hind lobe of prothorax black; superior appendage with apex excised at 90°
       (similar to Figs. PL9 or 12) ...._**chiriquita**_

Females

1   Propleuron chiefly black or dark brown **2**

1'   Propleuron pale blue or yellowish; apex of wings uncolored or fainty smoky **6**

2   Abdominal segment 9 brown or black, with a pair of blue or yellow spots which may or may not be confluent on the mid-dorsal line; 8 brown or black, no pale marks on dorsum **3**

2'   Abdominal segments 8 to 10 obscure, blackish, unmarked, or with only a trace of obscure bluish on the middle of each side of each segment in dorsal view; no pale humeral or antehumeral stripe or spot ...._**gigantula**_

3   Antehumeral stripe reduced to a small anterior spot ...._**dentata**_

3'   Antehumeral stripe present, not reduced to a spot **4**

4   Metepimeron with a dark stripe dorsal to the latero-ventral carina **5**

4'   Metepimeron without a dark stripe other than that on the metapleural suture
       ...._**distadens & baltodanoi**_

5   No dark stripe on the metasternum; blue antehumeral stripe 0.22-0.35 mm wide at midheight ...._**paulina & paulirica**_

5'   Dark stripe dorsal to the latero-ventral carina continued across the carina onto the metasternum ...._**angelina**_

6   Antehumeral stripe absent ...._**collaris**_

6   Antehumeral stripe present **7**

7   Abdominal segments 8 and 10 obscure brownish or blackish, 9 with a blue spot, usually longer than half of the segment, on each side of the dorsum; synthoracic dorsum brown or black with some metallic reflection; metepimeral-metasternal dark stripe absent ...._**nathalia**_

7'   Abdominal segments 8 and 9 brown, each with a blue spot on each side of dorsum, longer than half of the segment; synthoracic dorsum dark shining green, no pale antehumeral stripe or line; no dark metepimeral-metasternal stripe
       ...._**melanota**_

Derived from: BROOKS (1989), CALVERT (1931), DONNELLY (1992)

Protoneuridae

Plate VIII, p. 43

Small sized damselflies (30-35 mm abdomen length) with a very slender abdomen. Males are conspicuously colored with orange, yellow, red, or blue, whereas the females are generally brownish. They live in streams and swamps in primary and secondary forests, and can be regionally abundant. Males often hover low over the water surface for long periods, occasionally resting on leaf tips. Pairs oviposit in accumulations of floating debris (DUNKLE, in litt.).

Key to the genera of Protoneuridae

1	A not reduced to Ac, but extending beyond (Fig. PR1)	*Neoneura* (p. 46)
1'	Vein A reduced to Ac, not extending beyond	2
2	R4+5 and IR3 arising separately and the latter arising at the subnodus (Fig. PR2)	3
2'	R4+5 and IR3 joined at their origin into a single vein arising beyond the subnodus (sn)(Fig. PR3); or they may arise as separate veins, but if so, IR3 still arises beyond the subnodus	*Psaironeura* (p. 48)
3	Wings narrow, about one-sixth or one-seventh as wide as long; first antenodal space twice as long as the second and also longer than the third; Ac often very near the level of the first antenodals	*Protoneura* (p. 47)
3'	Wings broader, about one-fifth as wide as long; first antenodal space twice as long as the second, and equal to the third; Ac usually at a level intermediate between the antenodals (Fig. PR4)	*Epipleoneura* (p. 46)

Derived from: DONNELLY (1992), MUNZ (1919)

Epiploneura

Epiploneura letitia

This is a small (abdomen 26-29 mm long), inconspicuous protoneurid damselfly. Metathorax pale cream with dark brown line on posterior 2/3 of metepisternum, along second lateral suture; posterior dark spot on metepimeron; metallic green on dorsal half of metepisternum and posterior half of mesepimeron. Abdomen dark, traces of pruinescence on dorsum of segments 7-10. It flies „...close to quiet water on the margins of wooded streams." (DONNELLY, 1992).

Neoneura

Both in venation and general appearance, e.g. the more robust stature and the thoracic coloration, members of this genus resemble coenagrionids more than do some of the more extremely specialized Protoneuridae. They inhabit almost exclusively lotic waters.

Adults are generally found on and around floating or emergent vegetation around the larval habitat (WESTFALL & MAY, 1996).

Males

1	Thorax with a series of dark spots or a narrow stripe or bar along the mid-dorsal carina	**2**
1'	No mid-dorsal thoracic dark markings (except sometimes near the ante-alar sinus) or mesepimeral median dark stripe; abdomen with black predominating, eighth segment entirely dark above	***paya***
2	Abdomen predominantly pale	**3**
2'	Abdomen predominantly black	**4**
3	Abdomen largely red, dorsum of 8th and 9th segments largely pale	***esthera***
3'	Abdomen dull, pale; dorsum of 8th and 9th segments black	***aaroni*** (juv.)
4	Abdominal segment 2 pale above, bordered and patterned with black	***amelia***
4'	Segment 2 entirely dark above	***aaroni*** (adult)

Females

1	Lateral lobes of hind border of prothorax low, scarcely evident	***aaroni***
1'	Lateral lobes of hind border of prothorax conspicuous but not as high as middle lobe; abdominal segment 9 largely dark above	***amelia***
1''	Lateral lobes of hind border of prothorax conspicuous, about as high as middle lobe; segment 9 light yellow or brown above	***esthera***

After: WILLIAMSON (1917)

Protoneura

Males of this genus are recognizable by their remarkably attenuated abdomen and the brightly colored thorax contrasting to the dark colored abdomen. Besides the slender stature they are characterized by an extremely reduced venation, lacking any trace of the anal vein beyond Ac (WESTFALL & MAY, 1996).

They usually inhabitat the more lotic areas of small streams. Males are almost unvisible when flying in the shade, but when they come out in the sunlight „... to flutter almost motionless near the water´s surface, the red thoracic spots would suddenly appear like tiny flames..." (WILLIAMSON, 1915, about *P. calverti*). Females oviposit in submerged leaves.

1	CuP ending a little beyond the crossvein descending from the nodus	**2**
1'	CuP ending at the crossvein descending from the nodus; thoracic dorsum chiefly pale blue (male) or metallic green (female); abdomen violet-black or dark metallic green with a very narrow transverse basal pale ring on segments 3-7	***corculum***

2   R3 on hind wings arising nearest the fifth postnodal; males with superior appendages transversely bifid, inferiors longer; females with mesostigmal process bifid..*peramans*

2'   R3 on hind wings arising nearest the fourth postnodal; male superior appendages not bifid, female mesostigmal process simple...**3**

3   Inferior appendages of male with an acute superior tooth (Fig. PR6); female with a simple, acute, mesostigmal spine..**4**

3'   Inferior appendages of male with no acute superior tooth (Fig. PR5); female with a transversely elongated mesostigmal lamina; thoracic dorsum predominantly metallic green (mixed with brown in female), with orange-red (male) or yellow (female) median and double humeral lines.......................................*cara*

4   Thoracic dorsum predominantly black...**5**

4'   Thoracic dorsum predominantly pale...**6**

5   Thoracic dorsum predominantly black, each side with a pale blue antehumeral stripe (male) or a yellow posthumeral line (female), the latter sex also with a yellow median line...*cupida*

5'   Thoracic dorsum predominantly black, an orange antehumeral stripe, reaching upward three-fifths' to three-fourths' way toward the base of forewing (male); abdominal segments 3-6 of male chiefly orange-red, black at apices and along sides...*amatoria*

6   Thoracic dorsum predominantly orange, with a median black stripe (male); abdominal segment 3 vividly marked by orange.............................*aurantiaca*

6'   Thoracic dorsum predominantly vividly sulfur-yellow; abdominal segment 3 almost entirely black...*sulfurata*

Derived from: CALVERT (1908), DONNELLY (1989b)

Psaironeura

Small sized damselflies. Mesepisternum and most of mesepimeron dark metallic green, metepimeron pale yellow. Abdomen pale brown to brown with segment 9 or 9 and 10 pruinose bluish, the area which makes the insects visible to the observer when they „float" in the gloom of the forest (WILLIAMSON, 1915).

Both species are found in shaded swamps and grass-filled ponds in primary forests. They usually perch low over ground and remain in shade for most of the time (ESQUIVEL, 1994).

1   Labrum and clypeus of male brown (pale or dark); ventral projection of superior appendage, in lateral view, obtusely rounded (Fig. PR7); Ovipositor projecting backward as far as the tip of superior appds...*remissa*

1'   Labrum and clypeus of male bright red; ventral projection of superior appendage, in lateral view, narrow, parallel sided (Fig. PR8); Ovipositor projecting beyond the tip of superior appds..*selvatica*

After: ESQUIVEL (1994)

Coenagrionidae

Plate IX, p. 51

This is the largest family of Zygoptera. Its members can be found at a variety of habitats, from rivers to swamps, and sizes range between 20 and 45 mm length of the abdomen. They rest with abdomen usually held nearly horizontal. Males are often colored with blue, red, or yellow, whereas females are more cryptically colored.

Key to the genera of Coenagrionidae

1	Spines on the tibiae each much longer than the interval separating them	**2**
1'	Spines on the tibiae short (each one usually equal to or shorter than the interval separating it from its next neighbor)	**3**
2	Vein R3 arising at the fifth postnodal or further distad on the hind wing, at the sixth or further distad on the forewing; spines on the legs usually numbering more than 7 on the third tibiae; females without an apical ventral spine on abdominal segment 8	*Argia* (p. 53)
2'	Vein R3 arising at the fourth postnodal on the forewing (Fig. C1); spines on the legs fewer in number, 5-7 on the third tibiae; females usually with an apical ventral spine on abdominal segment 8	*Nehalennia* (p. 63)
3	R3 on the hind wings arising at or near to the fourth postnodal or further distad (Fig. C2)	**4**
3'	R3 on the hind wings arising at or near to the third postnodal or still nearer to the wing base (Fig. C9 & C10)	**11**
3''	R3 arising nearest to the fifth or sixth postnodal (Fig. C2)	*Metaleptobasis* (p. 62)
4	Costal side of the pterostigma on the hind wings (and often also on the forewings) usually shorter than the proximal or distal sides (Figs. C3 & 4)	**5**
4'	Costal side of the pterostigma on all the wings usually as long as or longer than the proximal or distal sides	**6**
5	A1 arising at least as far in front of cu-a as the latter is long (Fig. C3)	*Apanisagrion* (p. 53)
5'	A1 arising at or beyond cu-a (Fig. C4)	*Anisagrion* (p. 52)
6	Pterostigma bicolored or tricolored, usually dark posteriorly and pale anteriorly (wing venation as in Fig. C5)	*Enacantha* (p. 59)
6'	Pterostigma uniformly colored	**7**
7	Abdomen of male chiefly red or orange, almost unmarked; frons angulate in profile, with distinct transverse ridge at junction of antefrons and postfrons; generally lacking any postocular spots; female without an apical ventral spine on abdominal segment 8, ovipositor not extending beyond tip of abdomen	*Telebasis* (p. 63)
7'	No such combination of characters	**8**
8	Abdominal segments 8-10 predominantly blue, blue-green, or tan (in females), and variably marked with black; pale postocular spots on dark ground present; ovipositor of female extending barely beyond tip of abdomen at most	**9**

8'   Male abdominal segments 8-10 predominantly red, orange or yellowish with
        very little black; if postocular spots on dark ground present, then confluent with
        pale colouring of rear of head; ovipositior of female extending well beyond tip of
        abdomen..**10**

9   A1 arising at least as far in front of the submedian crossvein (cu-a) as the latter is
        long (Fig. C6); males with the hind margin of segment 10 more or less
        emarginate, but not usually elevated into a distinct tubercle or process
        ....*Enallagma* (p. 59)

9'   A1 arising at or slightly beyond the submedian crossvein (Fig. C7); males with
        the hind margin of segment 10 more or less elevated into a process or plate
        which is excised or bifid and projecting posterodorsally.....*Acanthagrion* (p. 50)

10   A1 extending only 5-6 cells, ending at a level between 1st and 2nd postnodals;
        dorsum of abdominal segments 8-10 yellow, segment 10 with forked dorsal
        projection; superior appendages of male straight (Fig. C13); female without
        ventral spine on segment 8 (Fig. C13)..............................*Chrysobasis* (p. 59)

10'   A1 in hind wing at least 7 cells long and ending at level of 3rd or 4th postnodal
        (Fig. C8); dorsum of abdominal segments 8-10 red, without a forked dorsal
        projection; superior appendages of male elbowed and bent down in their apical
        half (Fig. C11 & 12); female with or without ventral spine on segment 8 (Figs.
        C11 & 12)..*Leptobasis* (p. 62)

11   Quadrilateral cell in hind wing with anterior side about 2/3 as long as posterior
        side or longer (except for *I. ramburii*)(Fig. C9).......................*Ischnura* (p. 60)

11'   Quadrilateral cell in hind wing with anterior side not more than half as long as
        posterior side (Fig. C10)..................................*Neoerythromma* (p. 63)

Compiled from: CALVERT (1908), DONNELLY (1967), DONNELLY & ALAYO (1966),
MICHALSKI (1988), WESTFALL & MAY (1996)

Acanthagrion
Plate X, p. 55

In general appearance species of this genus are similar to *Enallagma*, with dark
markings strongly contrasting against a light (mostly light blue) background. They are
readily separated from other Neotropical genera by their declivent arrangement of the
male superior abdominal appendages. The genus *Anisagrion*, which is similar in this
respect, is separated from *Acanthagrion* by the lack of postocular spots. For
identification of the males it is necessary to dissect the accessory genitalia for a close
examination of the penis. For this purpose it is best to conserve the specimens in alcohol
rather than to dry them.

Little is known about the ecology of *Acanthagrion* species. It is believed that they
replace, in many ways, the holarctic genus *Enallagma* in the Neotropical region
(LEONHARD, 1977).

Plate IX

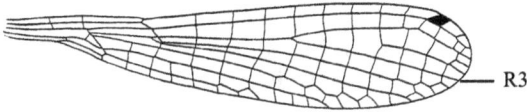

Fig. C1: *Nehalennia irene*, fore wing

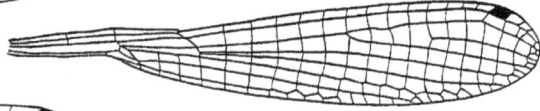

Fig. C2: *Metaleptobasis manicaria*, fore wing

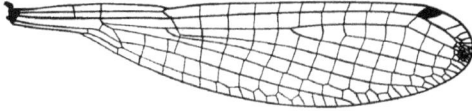

Fig. C3: *Apanisagrion lais*, hind wing

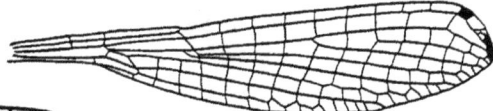

Fig. C4: *Anisagrion truncatipenne*, hind wing

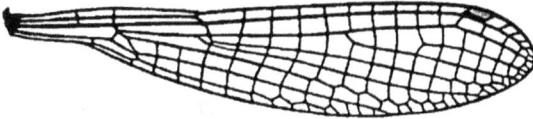

Fig. C5: *Enacantha caribbea*, hind wing

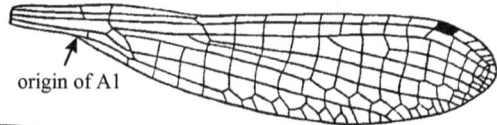

Fig. C6: *Enallagma civile*, hind wing

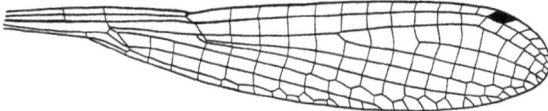

Fig. C7: *Acanthagrion gracile*, fore wing

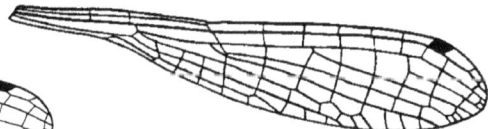

Fig. C8: *Leptobasis vacillans*, hind wing

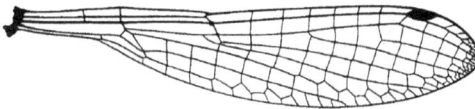

Fig. C9: *Ischnura capreolus*, hind wing

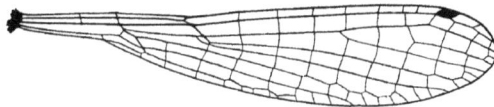

Fig. C10: *Neoerythromma cultellatum*, hind wing

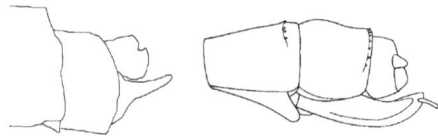

Fig. C11: *Leptobasis candelaria*, apex of abdomen of male (left) and female (right) in lateral view

Fig. C12: *Leptobasis vacillans*, apex of abdomen of male (left) and female (right) in lateral view

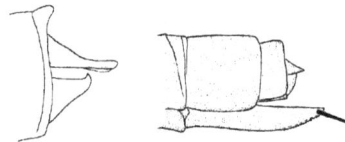

Fig. C13: *Chrysobasis lucifer*, apex of abdomen of male (left) and female (right) in lateral view

Males (Females very difficult to differentiate)

1   Ental surface of distal penis segment variously modified, but not armed with long, curved, sclerotized hooks; abdominal segment 10 higher than 9 (Fig. C20) **2**

1'   Ental surface of distal penis segment bearing a pair of long, curved, sclerotized hooks; abdominal segment 10 little if any higher than 9 in profile (Fig. C21) ....*inexpectum*

2   Modifications of distal penis segment, mesal or apical, extending laterad beyond margins of segment 2 in ventral aspect (Fig. C18) **3**

2'   Distal penis segment narrow basomesally; apically expanded to about twice its mesal width but not extending to margins of segment 2 (Fig. C19); when viewed dorsally, distal part of penis nearly parallel sided **4**

3   Prominent lateral lobes produced from distal penis segment between middle and base (Fig. C14); dorsum of abdominal segment 10 much elevated and constricted ....*trilobatum*

3'   Apex of distal penis segment greatly expanded, twice as wide as the portion of segment 2 which it overlies, lateral lobes not attaining margins of segment 2 (Fig. C15); dorsum of abdominal segment 10 moderately elevated, not constricted ....*quadratum*

4   Ental surface of penis bears a pair of fenestrate spurs which are straight or curved (Fig. C16) ....*kennedii*

4'   Ental surface of penis with a pair of small chitinized (darkened!) curved spurs (Fig. C17) ....*speculum*

Derived from: GARRISON (1985), LEONHARD (1977)

Anisagrion

Plate X, p. 55

Small sized damselflies similar to *Apanisagrion*, immmature males yellowish red to brownish, becoming blackish with age. Males with a bifid, dorsoapical prominence on abdominal segment 10. They usually are found at small rivulets.

1   Tip of hind wing more rounded than tip of fore wing, apex of abdomen as in Fig. C23 ....*truncatipenne*

1'   Wing tips in fore and hind wing equal **2**

2   Wings distal from the pterostigma light milky white, apex of abdomen as in Fig. C22 ....*kennedyi*

2'   Wings entirely hyaline ....*allopterum*

After: ST. QUENTIN (1960)

Apanisagrion

Apanisagrion lais

Males are easily distinguished from other coenagrionids, except *Anisagrion*, by a patch of dense venation at the tip of the hindwings, and by a sharp bending of R1 away from the wing margin at the pterostigma in the hind wings. Immatures are largely orange and might be mistaken for a *Telebasis* or a female *Ischnura*, but with age they become very dark with yellow green markings. Segments 8-10 becoming white pruinose dorsally, especially on 8 and 9. Dorsoapical margin of segment 10 slightly notched, not produced (WESTFALL & MAY, 1996). They usually inhabit grassy seepage areas (DUNKLE, in litt.).

Argia

Plates X (p. 55) & XI (p. 57)

These are medium to large sized damselflies which can be easily distinguished from other genera by the above listed characters. In addition, males uniquely have a specialized pair of pad-like structures on the dorsum of segment 10. They are usually found at streams with a significant current, perching on sunny places such as rocks. They commonly nervously flit their wings while perching, and move from perch to perch with a fast direct flight (DUNKLE, in litt.). A few species live in lakes or ponds as well (WESTFALL & MAY, 1996).

This is a very large Neotropical genus, comprising about 110 species and surely many still undescribed ones (GARRISON, 1994). Until now, only the North American species are studied well enough for a secure identification by GARRISON's work. The key for the Central American species provided by Calvert at the beginning of the 20th century is still the only one available and largely provisional.

Males (excl. *A. eliptica* and *A. funcki*)
Abdominal appendages as in Figs. C24 & 25

1	Thoracic dorsum not brilliantly metallic	**2**
1'	Thoracic dorsum brilliant metallic copper	**15**
2	Total area of dark colours on abdominal segments 3-6 and thoracic dorsum greater than the total pale area on the same parts	**3**
2'	Total area of dark colours on abdominal segments 3-6 and thoracic dorsum less than the pale area on the same parts	**17**
3	Labrum pale	**4**
3'	Labrum black; pale colours on segments 3-7 consisting only of a transverse ring, and a mid-dorsal stripe tapering posteriorly on 3 and a small mid-dorsal basal spot on 4	***rogersi***
4	Pale colours on dorsum of segments 3-6 limited to a transverse basal ring and at the most a fine mid-dorsal line; rear of head chiefly black	**5**
4'	Pale colours on dorsum of segments 3-6 consisting of a transverse basal ring, and a mid-dorsal stripe on some or all of them, tapering posteriorly; 8 mostly blue on dorsum (except in some indicatrix); rear of head chiefly black	**10**

5	Segment 8 mostly black on dorsum	**6**
5'	Segments 8 and 9 pale on dorsum	**8**
6	Segment 9 mostly pale on dorsum	*calida*
6'	Segment 9 mostly black on dorsum	**7**
7	Superior appendages not bifid at tip, basal half to fourth of dorsum of 9 pale	*translata*
7'	Superior appendages bifid at tip, apical third (or only a pair of apical spots) of 9 pale	*tezpi*
8	Segment 3 with basal half entirely blue, remainder of dorsum black, 8 and 9 with an inferior longitudinal black stripe each side; antenodal cells 5-4 on the forewings, 4 or 3+ on the hind; inferior appendages slightly bilobed in profile view	*terira*
8'	Segment 3 with basal sixth or less blue, remainder of dorsum black; 3 antenodal cells on the hind wings	**9**
9	Pale colours of thorax and base of abdomen blue, no inferior black stripes on segments 8 and 9, more often 4 antenodal cells on the forewings	*gaumeri*
9'	Pale colours of thorax and base of abdomen dark violet (nearly black when cool), segments 8 and 9 with an inferior longitudinal black stripe each side; more often 3 antenodal cells on the forewings	*frequentula & pulla*
10	Superior appendages distinctly bifid, inner branch much longer than the outer	**11**
10'	Superior appendages variously formed, but not distinctly bifid; no apical, dorsal, black spot on 8	**12**
11	Segment 8 with no apical dorsal black spot; mid-dorsal pale stripe on 3 and 4 only	*ulmeca*
11'	Segment 8 with an apical dorsal black spot; mid-dorsal pale stripe on 3-6	*adamsi*
12	Antenodal cells on the forewings usually more than 3	*difficilis & herberti & indicatrix & oculata & popoluca*
12'	Antenodal cells on the forewing usually 3	**13**
13	Mid-dorsal thoracic black stripe at least a little wider than either pale antehumeral stripe; rear of head chiefly black	**14**
13'	Mid-dorsal thoracic black stripe reduced to a line; rear of head chiefly pale; inferior appendages a little longer than the superiors; second lateral thoracic suture with a black stripe for its entire length	*talamanca*
14	Inferior appendages a little shorter than the superiors; second lateral thoracic suture with a black stripe for its entire length	*underwoodi*
14'	Inferior appendages twice as long as the superiors; second lateral thoracic suture with a black mark at its upper end only	*johanella*
15	Labrum chiefly metallic copper, at least in its basal half, apical half for only its front edge yellow; lower branch of inferior appendage less robust than the upper branch	**16**
15'	Labrum yellow or orange throughout; lower branch of the inferior appendages more robust than the upper branch	*oenea*
16	Segments 3-8 black on dorsum, with only a narrow, basal, blue ring; 9-10 blue on dorsum	*cuprea*
16'	Segments 3-6 blue on dorsum, apical sixth to fifth black; 7 black with basal blue ring; 8-10 blue on dorsum	*cupraurea*

Plate X

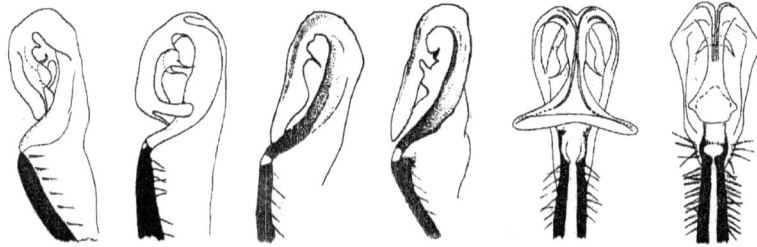

C14	C15	C16	C17	C18	C19
A. trilobatum	*A. quadratum*	*A. kennedii*	*A. speculum*	*A. quadratum*	*A. kennedii*

Fig. C14-17: Penis of *Acanthagrion*, lateral view

Fig. C18 & 19: Penis of *Acanthagrion*, ventral view

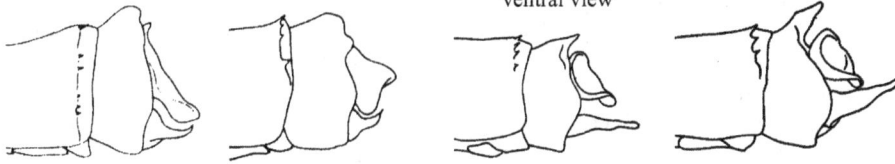

C20: *A. kennedii*　　**C21**: *A. inexpectum*　　　**C22**: *A. kennedyi*　　**C23**: *A. truncatipenne*

Fig. C20 & 21: *Acanthagrion* males, apices of abdomen in lateral view

Fig. C22 & 23: *Anisagrion* males, apices of abdomen in lateral view

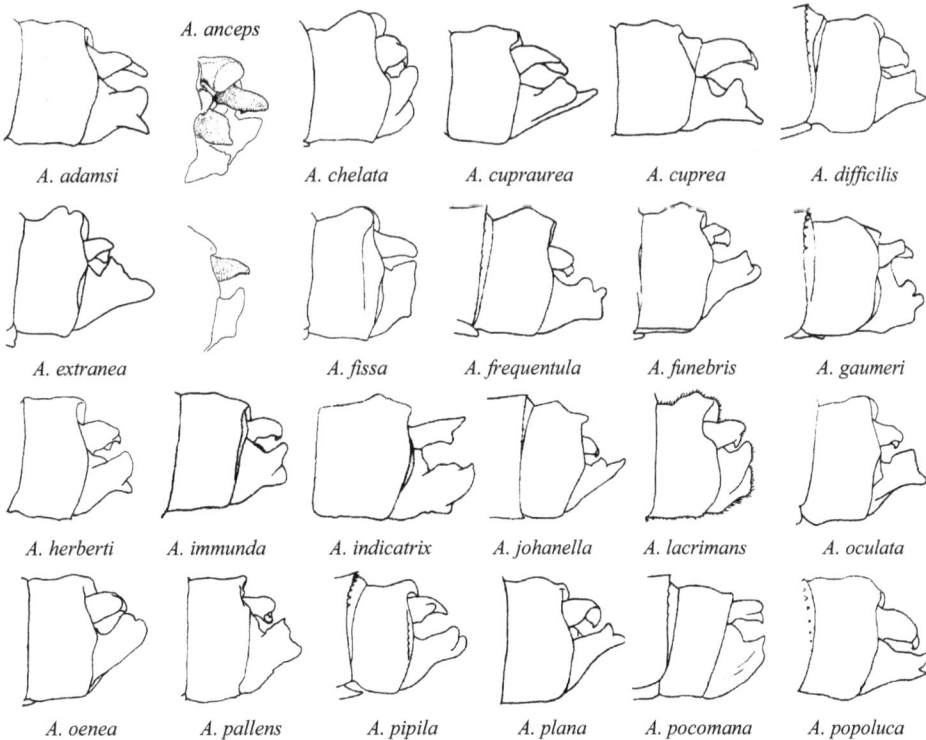

A. adamsi　　*A. anceps*　　*A. chelata*　　*A. cupraurea*　　*A. cuprea*　　*A. difficilis*

A. extranea　　*A. fissa*　　*A. frequentula*　　*A. funebris*　　*A. gaumeri*

A. herberti　　*A. immunda*　　*A. indicatrix*　　*A. johanella*　　*A. lacrimans*　　*A. oculata*

A. oenea　　*A. pallens*　　*A. pipila*　　*A. plana*　　*A. pocomana*　　*A. popoluca*

Fig. C24: *Argia* males, abdominal appendages in lateral view

17   Rear of head chiefly black; postbasal black streaks on 3-6; 8 and 9 blue, with an
        inferior black stripe each side; superior appendages distinctly bifid, inner branch
        longer..**18**

17'   Rear of head pale; no postbasal black stripes on 3-6; 8 and 9 blue, almost
        unspotted...*anceps & fissa & lacrimans & westfalli*

17''   Rear of head largely or chiefly black; dorsum of segment 7 blue for its entire
        length; a continuous inferior black stripe on each side of 3-9 for almost their
        entire length; superior appendages not bifid at tip ....*variabilis*

17'''   Rear of head pale; dorsum of segment 7 chiefly black (or if blue then with black
        postbasal stripes on 3-6, or an inferior black stripe on each side of 8 and 9, or
        with both, hence different from 17') with or without black postbasal streaks on
        3-6; 8-10 at least blue on dorsum; superior appendages variable, but when bifid
        the branches are subequal...**19**

18   Inferior appendages longer than high; inferior appendages bifid at tip; lower
        branch of inferior appendages subequal in length to the upper branch; black
        humeral stripe narrower in its upper half...*pipila*

18'   Inferior appendages higher than long...*chelata*

19   Superior appendages almost entire on tip, the inner margin with a small,
        subacute, anteapical projection..**20**

19'   Superior appendages bifid at apex, branches subequal in length; 3 antenodal cells
        in the forewings; postbasal black streaks present and continuous with the apical
        black on 3-7; 8-10 blue with an inferior longitudinal black stripe each side

        ....*pocomana*

19''   Superior appendages trilobed or bilobed at apex, inner margin rounded and
        convex before the apex; antenodal cells on the forewings 4....................*pallens*

20   Antenodal cells on forewings 4...**21**

20'   Antenodal cells on forewings 3; postbasal black stripes present on segments 3-6;
        8 and 9 pale, each side with an inferior black stripe.................................*immunda*

21   Inferior appendages bifid at tip, lower branch distinctly longer than the upper;
        black postbasal streaks present on 3-6, usually not confluent with the apical
        black...*extranea*

21'   Inferior appendages bifid at tip, lower branch equal to or shorter than the upper

        ....**22**

22   Black postbasal streaks present on segments 3-6..................................*plana*

22'   Black postbasal streaks absent from segments 3-6...............................*funebris*

Females

1   Dorsum of abdominal segments 3-6 mostly black...**2**

1'   Dorsum of abdominal segments 3-6 chiefly pale...**15**

2   Dorsum of segments 8 and 9 pale, with black markings.............................**3**

2'   Dorsum of segment 8 black, a pale blue spot either side at base; 9 blue on
        dorsum; no mesepisternal tubercles...*popoluca*

2''   Dorsum of 8 blue, of 9 black, with apical half (or a pair of apical spots) blue; a
        pale mid-dorsal stripe on 3 and 4 at least, 10 black; mesepisternal tubercles
        present...*variabilis*

2'''   Dorsum of segments 8 and 9 pale, unspotted (except for a pair of basal dorsal
        black dots on 9 in pocomana)...**12**

Plate XI

A. pulla *A. rogersi* *A. talamanca* *A. terira*

A. tezpi *A. translata* *A. ulmeca* *A. underwoodi* *A. variabilis*

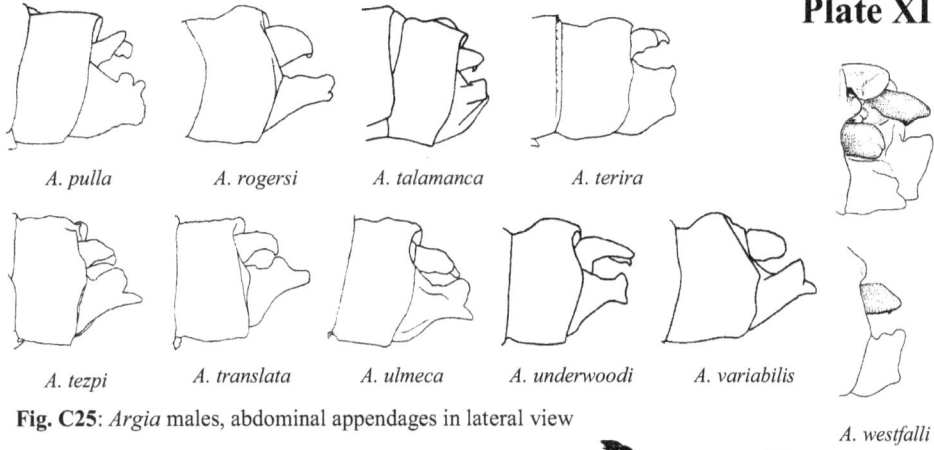

Fig. C25: *Argia* males, abdominal appendages in lateral view

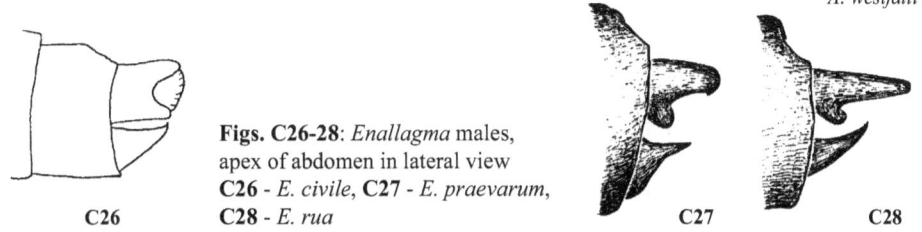

A. westfalli

Figs. C26-28: *Enallagma* males, apex of abdomen in lateral view
C26 - *E. civile*, **C27** - *E. praevarum*,
C28 - *E. rua*

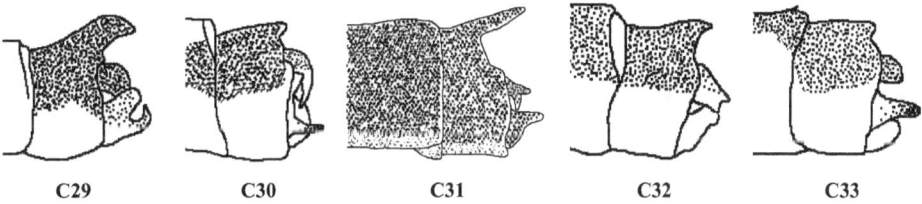

C26   C27   C28

C29   C30   C31   C32   C33

Figs. C29-33: *Ischnura* males, apex of abdomen in lateral view **C29**: *I. capreola*; **C30**: *I. denticollis*; **C31**: *I. hastata*; **C32**: *I. posita*; **C33**: *I. ramburii*

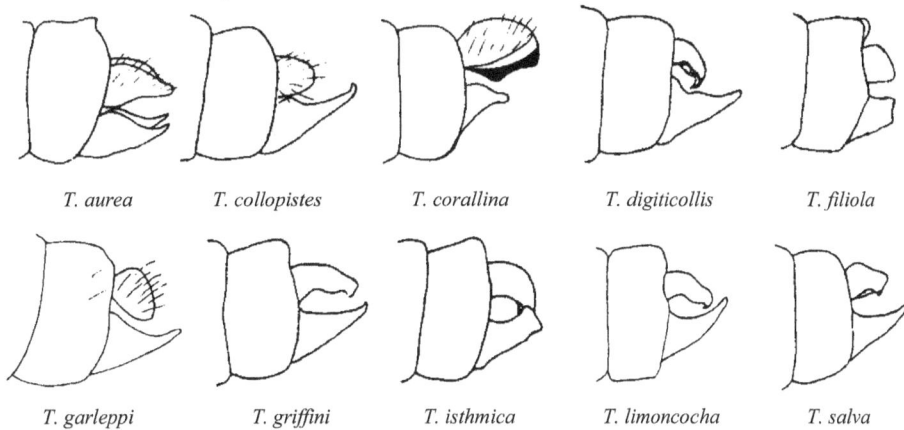

T. aurea *T. collopistes* *T. corallina* *T. digiticollis* *T. filiola*

T. garleppi *T. griffini* *T. isthmica* *T. limoncocha* *T. salva*

Fig. C34: *Telebasis* males, lateral view of abdominal appendages

3	These markings consisting of two stripes occupying only the basal half of segments 8 and 9	**4**
3'	These markings consisting of two longitudinal stripes, more or less confluent at base, occupying the basal fourth or less of segment 8 and nearly the whole length of 9; mesepisternal tubercles well developed; antenodal cells 5-4 on forewings, 4-3+ on hind wings	*terira*
3''	These markings consisting of two stripes as long as segment 8 and nearly as long as 9; 5 antenodal cells on the forewings, 4 on the hind	**8**
4	Mesepisternal tubercles present but small; segment 10 pale on dorsum	**5**
4'	Mesepisternal tubercles absent; 3 antenodal cells on the hind wings	**6**
5	Labrum pale; a pale mid-dorsal stripe on segments 3 and 4, or on 3 only; 4 antenodal cells on the hind wings	*ulmeca*
5'	Labrum black; a pale mid-dorsal stripe on segments 3-5; 3 antenodal cells on the hind wings	*adamsi*
6	A pale mid-dorsal stripe on segments 3-5, 10 black on dorsum, with a pair of small pale spots	*oculata*
6'	A pale mid-dorsal stripe on segment 3 only, or absent altogether	**7**
7	Segment 10 pale on dorsum; labrum black; wings clear	*difficilis*
7'	Segment 10 black on dorsum; labrum pale; wings yellow or smoky	*indicatrix*
8	Mesepisternal tubercles well developed	**9**
8'	Mesepisternal tubercles absent or very small	**11**
9	Segment 10 pale, whith a pair of brown or black spots on dorsum	*translata*
9'	Segment 10 pale, unspotted on dorsum	**10**
10	Thoracic dorsum with no coppery-red reflection	*tezpi*
10'	Thoracic dorsum, and usually the vertex, nasus, and labrum also, with a coppery-red reflection	*cuprea & cupraurea*
11	Mesepimeral black stripe reaching upward almost to, or to, the upper mesepimeral margin; dark colours of head and thorax with some metallic reflection	*oenea*
11'	Mesepimeral black stripe reaching upward only half-way to the upper mesepimeral margin; dark colours of head and thorax without metallic reflection	*pipila*
12	Antenodal cells in forewing more often 4 or 3+; no pale mid-dorsal longitudinal stripe on segment 3	*gaumeri*
12'	Antenodal cells on forewing more often 3	**13**
13	Segment 9 with a pair of basal dorsal black dots; rear of head pale; a mid-dorsal longitudinal pale stripe on segments 3-5	*pocomana*
13'	Segment 9 with no black marks	**14**
14	Second lateral thoracic suture with a black stripe on its entire length; pale mark on each side of rear of head elongated transversely	*underwoodi*
14'	Second lateral thoracic suture with a black mark only at its upper end; pale mark on each side of rear of head elongated vertically	*frequentula & johanella & pulla*
15	Postbasal black stripes present on segments 3-6, and confluent with the apical black (some specimens of extranea, ?variabilis, ?pallens and ?plana will fall here, instead of under 15')	*lacrimans*
15'	Postbasal black stripes present on segments 3- or 4-6, but not usually confluent with the apical black	**16**

15''   Postbasal black streaks absent on segments 3-6, apical black reduced to a spot each side; 8-10 pale, unspotted..***fissa***

16   Segments 8 and 9 pale, without black on dorsum (except on 9 in many specimens of extranea and ?plana)..**17**

16'   Segments 8 and 9 pale, each with a longitudinal black stripe on each side of dorsum, from the base backward to a variable distance...........................***pallens***

17   Antenodal cells on forewings 4...**18**

17'   Antenodal cells on forewings 3; mid -dorsal thoracic stripe present; humeral stripe reduced to a line; no black marks on dorsum of segment 9; mesepisternal tubercles absent...***immunda***

18   Mesostigmal lamina rounded externally...***extranea***

18'   Mesostigmal lamina angulate externally...***plana***

Derived from: CALVERT (1908), GARRISON (1996)

Chrysobasis

Chrysobasis lucifer

This small sized damselfly inhabits small ponds in wooded or partially wooded areas. The bright yellow tip of the abdomen is the only conspicuous marking, which appears „... at first glance to be a detached spot of light moving slowly around in the shadows...“ (DONNELLY, 1967).

Enacantha

Enacantha caribbea

This species externally resembles very much an *Enallagma* with somewhat reduced black markings, but it is set apart distinctly by the bicolored pterostigma. It occurs in ponds in the eastern lowlands of southern Mexico, Guatemala and Belize (and Cuba) and can be locally abundant (WESTFALL & MAY, 1996).

Enallagma
Plate XI, p. 57

These are small to medium sized damselflies inhabiting a variety of habitats, ranging from ponds (*E. civile*) to streams (*E. novaehispaniae*), preferably in open areas. The color pattern of the male abdomen consists of dark markings on blue to violet background.

Males

1   Black markings on abdominal segment 2 not U-shaped, pale colour on 3 blue....**2**

1'   Black markings on abdominal segment 2 U-shaped, pale colour on 3 violet; superior appendages widely bifid in profile view, upper branch longer ....***novaehispaniae***

2   Hind margin of prothorax convex throughout; superior appendages bifid at tip, upper branch longer, a pale tubercle between it and the lower branch (Fig. C26); abd. segments 4 and 5 predominantly blue...***civile***

2'   Hind margin of prothorax convex medially, slightly concave or truncated on each side; superior appendages with a distinct ventrally directed process or lamina in their basal half...**3**

3   Superior appendage equal in length to 10th abdominal segment at mid-height, distinctly longer than high in lateral view (Fig. C28)..***rua***

3'   Superior appendage distinctly shorter than 10th abdominal segment (Fig. C27) ....***praevarum***

Females

1   Hind margin of prothorax convex throughout...**2**

1'   Hind margin of the prothorax convex medially, slightly concave or truncated on each side; black on dorsum of abd. seg. 1 not usually reaching to the apex, dorsum of 8 and 9 black, which is narrower anteriorly on 8 and posteriorly on 9 ....***rua & praevarum***

2   Black on dorsum of abd. seg. 1 not reaching the apex................***novaehispaniae***

2'   Black on dorsum of abd. seg. 1 reaching to the apex of the segment; dorsum of 8 and 9 black, of approximately equal width throughout.................................***civile***

Derived from: CALVERT (1908), DONNELLY (1968)

Ischnura

Plate XI, p. 57

Very small to medium sized damselflies. Coloration of dorsum of male abdomen variable, black with blue markings in *I. denticollis*, *I. capreola* and *I. ramburii*, without blue colors in *I. posita*, and bright yellow in *I. hastata*. They occur preferably in ponds and marshes and individuals usually remain in dense vegetation.

Males

1   Pterostigma of forewing separated from costa, abdominal segment 10 with a spine-like dorsoapical prominence about ½ as long as segment 9 (Fig. C31); abdomen mostly yellow dorsally..***hastata***

1'   Pterostigma of forewing bordered anteriorly by costa; abdominal segment 10 with dorsoapical prominence usually much shorter, not spine-like; abdomen not yellow dorsally...**2**

2   Arculus distal to second antenodal crossvein by a distance equal to its upper arm; dorsoapical prominence of abdominal segment 10 well developed, bifurcated for nearly ½ its length, and strongly curved posteriorly; inferior appendages strongly bifid, the inferior branch longer than the superior branch (Fig. C29).......***capreola***

2'   Arculus at the second antenodal or distal to it by much less than length of its upper arm; dorsoapical prominence of abdominal segment 10 not so, if curving slightly posteriorly then low and blunt; inferior appendages variable, but usually not as above **3**

3   Pale antehumeral stripes entirely absent, the mesepisterna solid black; apex of abdomen as in Fig. C30 ....*denticollis*

3'   Pale antehumeral stripes present although sometimes represented only by a small anterior and posterior spot on each side **4**

4   Superior appds. distinctly bifid, having a lateral posteriorly directed process and a medial ventrally directed process of nearly equal length (Fig. C33)......*ramburii*

4'   Superior appds. hooked downward and entire, with at most a small posteriorly directed tooth projecting from the main branch at about the point where the latter turns downward (Fig. C32); abdominal segment 8 largely blue on dorsum
        ....*posita acicularis*

Females

1   Arculus distal to second antenodal crossvein by a distance equal to its upper arm; middle lobe of prothorax with a very prominent, mound-like protuberance on each side; hind margin of prothorax with two small, lateral lobes and an abrupt median excavation, sometimes with a small, subtriangular projection at the center of the latter ....*capreola*

1'   Arculus at the second antenodal crossvein or distal to it by much less than length of its upper arm; middle lobe of prothorax without mound-like protuberance; hind lobe of prothorax not as above **2**

2   Middle lobe of prothorax with a definite tooth-like or nipple-like process on each side, often marked with pale color ....*denticollis*

2'   Middle lobe of prothorax without such processes, usually smoothly rounded.....**3**

3   No vulvar spine on abdominal segment 8; humeral stripe always present; antehumeral stripe usually divided into two parts so as to resemble an exclamation mark or at least constricted at about ¼ the distance from the posterior end, the pattern often obscured with age, but visible if thorax wetted with alcohol or acetone; postquadrangular antenodal cells usually 2...........*posita*

3'   Vulvar spine normally present on abdominal segment 8, often large; either with no dark humeral stripe or with the antehumeral stripe complete; postquadrangular antenodal cells 3 ....**4**

4   Middle lobe of pronotum without distinct pits, sometimes with pair of narrow transverse grooves; middorsal thoracic carina ending abruptly at anterior end in a narrow, transverse ridge that extends across gap between posteromedial corners of mesostigmal plates; hind wing 16 mm or longer.......................................*ramburii*

4'   A pair of small, well-defined pits on each side of middle lobe of pronotum; middorsal thoracic carina ending anteriorly in obtuse bifurcation slightly behind mesostigmal plates; hind wing 15 mm or shorter ....*hastata*

Derived from: DONNELLY (1965a), NOVELO GUTIÉRREZ & PEÑA OLMEDO (1989), WESTFALL & MAY (1996)

Leptobasis
Plate IX, p. 51

Small sized damselflies with a reddish abdomen and thus generally similar to *Telebasis*, but distinguished by a rounded frons, the presence of postocular spots, and a different thoracic color pattern typically as follows: pale blue green laterally, a black middorsal stripe 1/3 to 1/2 width of mesepisterna, a yellow antehumeral stripe 1/3 to slightly more than 1/2 width of middorsal stripe, a black humeral stripe 2/3 width of middorsal stripe or more, covering dorsal 2/3 of mesinfraepisternum. However, there are widely different color forms which are probably mainly a function of age (WESTFALL & MAY, 1996). Legs are pale yellowish, in both sexes with reduced supplementary tooth on the tarsal claw. Females are distinctive because of their long ovipositor, which extends as much as 1/4 of its length beyond the distal margin of segment 10.
Although PAULSON (1997) listed *L. candelaria* only for Mexico, according to a note of WESTFALL & MAY (1996) this species is also recorded from Guatemala and thus is included in the key.

Males
1   Inferior appendages not curving strongly inward when viewed dorsally, only slightly longer than superiors, the latter with a broad triangular projection, acute at the apex and directed ventromedially (Fig. C12)............................*vacillans*
1'   Inferior appendages curving strongly inward when viewed dorsally, and much longer than the superiors, the latter without a triangular projection (Fig. C11)
        ....*candelaria*

Females
1   Ovipositor extending as much as 1/4 its length (excl. styli) beyond the distal margin of abdominal segment 10 and with ventral margin straight for most of its length ; vulvar spine normally present (Fig. C12)............................*vacillans*
1'   Ovipositor not extending so far beyond the distal margin of abdominal segment 10 and with ventral margin distinctly convex; vulvar spine absent (Fig. C11)
        ....*candelaria*

After: WESTFALL & MAY (1996)

Metaleptobasis

Medium sized damselflies. Males of the two species occuring in Central America have an orange colored thorax and a dark metallic green (*M. bovilla*) or dark brown (*M. westfalli*), rather slender abdomen (CALVERT, 1908; CUMMING, 1954). No character is known so far to separate the females.
They can be found in stream backwaters and seepage areas where the adults fly slowly , phantom-like, between low perches (DUNKLE, in litt.).

Males

1   Mesothoracic horns stout, each subparallel with its fellow of the opposite side for half of its length, then strongly diverging and curving laterad to a rounded tip ....*bovilla*

1'   Mesothoracic horns slender, with no strong divergence at the apices and with sharp tips, gently converging throughout most of their length, quite close together at their tips; superior abdominal appendages in dorsal view with no lateral divergence at the tips...*westfalli*

After: CUMMING (1954)

Nehalennia

Nehalennia minuta

This is a very small, slender damselfly with the abdomen being largely black with slight metallic sheen except abdominal segments 9 and 10, which are entirely blue on dorsum. The thorax is blue with black on dorsum, and a black stripe on the mesepisternum. They live usually in rather open country at ponds where they are very inconspicuous not only because of their small size, but also because they rarely fly outside dense vegetation (WESTFALL & MAY, 1996).

Neoerythromma

Neoerythromma cultellatum

Males of this species are distinguished from other damselflies by their bright yellow face. In addition, they have a pair of blue postocular spots, a pair of yellow-olive frontal thoracic stripes, blue thorax sides, and blue on dorsum of abdominal segments 8 and 9. Females always lack a vulvar spine. Breeding adults are associated with floating or barely submerged vegetation in lentic waters (DUNKLE, 1990; WESTFALL & MAY, 1996).

Telebasis

Plate XI, p. 57

Males of these medium sized damselflies are characterized by their almost unmarked red abdomens. Females are much duller. Despite their conspicuous coloration they often are not easy to detect when resting among dense vegetation around ponds and marshes.

Males (Abdominal appendages as in **Fig. C34**)

1	Wings flavescent	*aurea*
1'	Wings hyaline	**2**
2	In lateral view, superior appendage shows an elongate seam, appearing two-parted	*corallina*
2'	Superior appendage without such a seam	**3**
3	Rear of head mostly black	**4**
3'	Rear of head mostly pale	**9**
4	Labrum blue or black	**5**
4'	Labrum red	**6**
5	In lateral view, superior appendage rounded; abdomen 35 mm	*garleppi*
5'	Superior appendage not rounded; abdomen 17-23 mm	*filiola*
6	In lateral view, superior appd. rounded; inferior appd. pointed	*collopistes*
6'	In lateral view, superior appd. not rounded	**7**
7	Superior and inferior appendage extend caudad to the same or almost the same level, apex of inferior appd. averages only 0.05 mm beyond that of the superior appd.; superior appd. not or only slightly bent ventrad	*griffini*
7'	Superior appendage extends caudad not as far as inferior, the latter averaging 0.1 mm beyond the superior appd.; superior appd. strongly bent ventrad	**8**
8	Posterior prothoracic lobe dorsally black; mesepimeron usually with a black stripe; first lateral suture without a black mark at mid length	*digiticollis*
8'	Posterior prothoracic lobe dorsally brown or tan; mesepimeron without a black stripe; first lateral suture often with a black mark at about mid length	*limoncocha*
9	Posterior portion of mesepisternal black with a distinct lateral projection	*salva*
9'	Posterior mesepisternum without such a black projection	**10**
10	In dorsal view, superior appd. much wider than long	*boomsmae*
10'	In dorsal view, superior appd. not wider than long	*isthmica*

After: BICK & BICK (1995)

Females

1	Prothoracic horns obvious, extending anteriorly from hind lobe	**2**
1'	These horns absent or minute and difficult to detect	**7**
2	Rear of head mostly black, carina dark	**3**
2'	Rear of head mostly pale, carina either dark or pale	**6**
3	Abdomen 35 mm, hind wing 24 mm	*garleppi*
3'	Abdomen 24 to 31 mm	**4**
4	Humeral suture with a distinct, narrow, black line; mesinfraepisternum with a C-shaped black mark	*aurea*
4'	Humeral suture and mesinfraepisternum not as above	**5**
5	Dorsal surface of middle prothoracic lobe almost entirely black	*digiticollis*
5'	Dorsal surface of middle prothoracic lobe mostly brown	*limoncocha*
6	Posterior part of mesepisternal black widened laterally	*salva*
6'	Posterior part of mesepisternal black not so widened	*corallina*
7	Rear of head half black, half pale	*griffini*

7'   Rear of head either mostly black or mostly pale **8**
8   Rear of head mostly black **9**
8'   Rear of head mostly pale **11**
9   Carina dark; abdomen 24-25 mm **10**
9'   Carina pale; abdomen 16-18 mm .... *filiola*
10   Anterior margin of the mid-dorsal carina with a small elevated projection
         .... *collopistes*
10'   Mid-dorsal carina without such a projection .... *griffini*
11   Carina very dark bronze or black .... *boomsmae*
11'   Carina pale .... *isthmica*

After: BICK & BICK (1996)

Aeshnidae

Plate XII, p. 67

Large sized dragonflies having very large eyes and long, slender abdomens. When resting, they hang down in a vertical position. These are strong fliers, which usually make long patrolling bouts and are not easy to catch. Many species are known to be crepuscular or flying within deep forests, and thus are rarely encountered in broad daylight. Many species are attracted by light traps or lighted buildings during evening hours, so this sometimes is a much more effective and easier way to get them.

Key to the genera of Aeshnidae

1	R3 forming a marked bend near distal end of pterostigma; male hindwings lacks anal angle and anal triangle (Fig. AE1)	*Anax* (p. 68)
1'	R3 forming a regular curve under pterostigma; male hindwing with anal angle and anal triangle	**2**
2	Triangles with proximal cell usually free, if a crossvein is present, it ends at anterior side of triangle (Fig. AE10)	**3**
2'	Triangles with proximal cell crossed, at least in forewing, its crossvein ending at first triangular crossvein (Fig. AE11)	**5**
3	IR3 forking proximal to level of pterostigma; males with anal triangle 3-celled (Fig. AE2)	*Aeshna* (p. 68)
3'	IR3 forking under pterostigma or at the level of its proximal end (Fig. AE3), males with anal triangle 2-celled	**4**
4	Rspl not reaching border of wing distally, becoming nearly indistinct at two rows of cells from IR3 (Fig. AE3); male hindwing with A3 joined with border of wing after the anal angle; female cerci short, about length of abdominal segment 10	*Remartinia* (p. 72)
4'	Rspl reaching the border of wing, distally separated by one row of cells from IR3 (Fig. AE4); male hindwing with A3 joined with border of wing before the anal angle; female cerci long, about length of abdominal segments 8-10 together or longer	*Coryphaeschna* (p. 69)
5	Supratriangle in hind wing longer than median cell	**6**
5'	Supratriangle in hind wing shorter than or as long as median cell	**9**
6	Two rows of cells between R2 and R3 beginning under the pterostigma or more distally in hind wing	**10**
6'	Two rows of cells between R2 and R3 beginning at the basal end of pterostigma or more basally in hind wing	**7**
7	Subcosta not prolonged beyond nodus (Fig. AE5)	*Gynacantha* (p. 70)
7'	Subcosta prolonged through and beyond nodus	**8**
8	Midbasal space reticulate; some crossveins proximal to primary antenodal	*Neuraeschna* (p. 71)
8'	Midbasal space free (or with a single crossvein); crossveins proximal to first primary antenodal absent	*Staurophlebia* (p. 72)
9	IR3 apparently unbranched; two crossveins under the pterostigma	*Oplonaeschna* (p. 72)

Plate XII

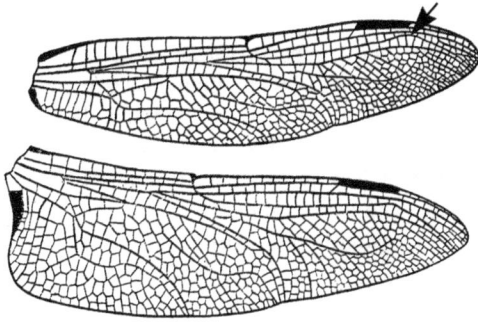

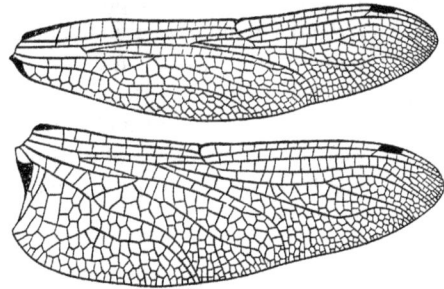

Fig. AE1: Wings of *Anax junius*, showing bent of R3 near distal end of pterostigma

Fig: AE2: Wings of *Aeschna psilus*

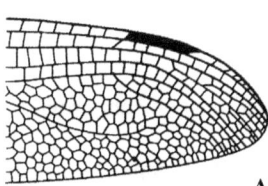

AE3

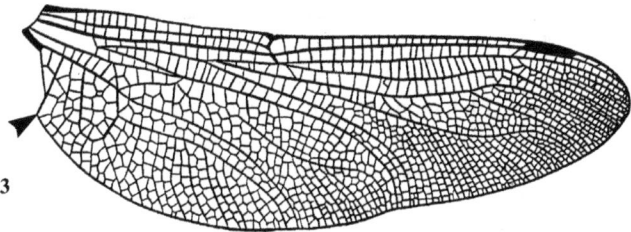

Fig. AE5: Hind wing of *Gynacantha nervosa*

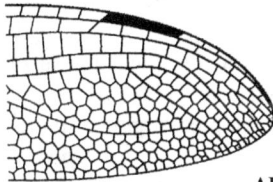

AE4

AE6

AE7

Figs. AE6 & 7: *Coryphaeschna diapyra* (AE6) and *C. amazonica* (AE7), base of abdomen of male in lateral view

Figs. AE3 & 4: Apex of right fore wing of *Remartinia luteipennis* (AE3) and *Coryphaeschna ingens* (AE4)

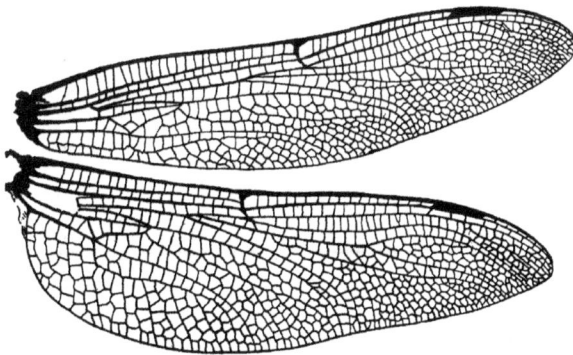

Fig. AE8: Wings of *Epiaeschna heros*

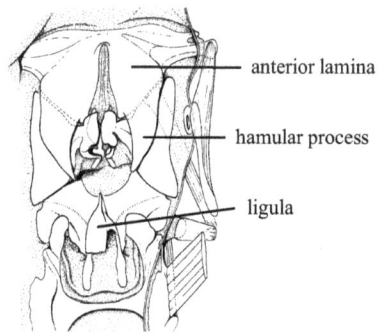

anterior lamina

hamular process

ligula

Fig. AE9: Male accessory genitalia (*Aeshna*), ventral view (vesica spermalis removed)

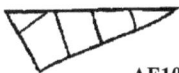

AE10

AE11

Figs. AE10 & 11: Triangles of *Coryphaeschna viriditas* (AE10) and *Gynacantha nervosa* (AE11), fore wing

9' IR3 distinctly branched; three or more crossveins under the pterostigma ....*Aeshna* (p. 68)

10 Fork of IR3 near the proximal end of the pterostigma in forewing (except in *trifida*) and more basal in the hind wing than in the forewing; many rows of cells between IR3 and Rspl at the level of the fork; R2 and R3 of hind wing parallel to the level of the pterostigma ....*Triacanthagyna* (p. 73)

10' Fork of IR3 in front and hind wing more or less at the same distance to the pterostigma; two rows of cells between IR3 and Rspl at the level of the fork; R2 and R3 of hind wing approximated at the pterostigma (Fig. AE8) ....*Epiaeschna* (p. 69)

Derived from: CARVALHO (1992), NEEDHAM (1929)

Aeshna

Typical aeshnid dragonflies with pale thoracic and abdominal markings on dark ground.

1 Males with superior appendages in lateral edge view not bifid at apex, nor with an anteapical inferior point; point of beginning of two rows of cells between R2 and R3 variable **2**

1' Males with superior appendages in lateral edge view bifid in apical fourth or less, the lower division much shorter than the upper; both sexes with two rows of cells between R2 and R3 beginning distal to the level of the pterostigma or under the distal end thereof ....*jalapensis*

2 Pale mesepimeral stripe almost straight in at least the lower 6/7 of its course; male superior appendages with no inferior subbasal tooth or tubercle, or this rudimentary ....*williamsoniana*

2' Pale mesepimeral stripe at least a little curved **3**

3 Anterior lamina of male (see Fig. AE9) with spines 0.5 - 0.8 mm long; female anal appendages 3.44-6.80 mm long ....*cornigera*

3' Anterior lamina of male with rudimentary spines 0.09 - 0.14 mm long; female anal appendages 7.28 - 8.26 mm long ....*psilus*

After: CALVERT (1956)

Anax

These are large sized dragonflies with an unmarked bright green thorax. Males have the abdomen black with more or less extensive blue markings. Males lack auricles on abdominal sgement 2 (DUNKLE, 1989; NEEDHAM, 1929).

1 Top of frons green, without dark markings ....*concolor*

1' Top of frons with a black spot **2**

2 Black spot on frons not enclosed by blue anteriorly ....*amazili*

2'   Black spot on frons surrounded with blue..**3**
3   Wingspan about 105 mm, superior appendages of male widest at three-fourth of
        their length, not bifid in profile view..***junius***
3'   Wingspan about 120 mm, superior appendages of male widest at apex, bifid in
        profile view..***walsinghami***

Derived from: CALVERT (1908), NEEDHAM (1929)

Epiaeschna

Epiaeschna heros

Very large sized dragonfly. Dark brown with blue eyes, green stripes on thorax, and
narrow green abdominal rings. They can be found on shady woodland ponds and small
slow streams, particularly in swampy areas (DUNKLE, 1989).

Coryphaeschna

Large sized dragonflies with long abdomens. Males either with thorax green with brown
stripes (sometimes reduced to narrow lines) and abdomen black with green rings, or
males entirely reddish without conspicuous markings (*C. amazonica*, *C. diapyra*, *C.
perrensi*). They are swift-flying, powerful insects often found around weedy lakes,
marshes, or mangrove swamps (DUNKLE, 1989). Males of some species are known to
form feeding swarms, especially at dusk.

1   Thorax with a conspicious large spot on the mesepisternum; inferior appendage
        of male reaches two thirds the length of the superiors.............................***apeora***
1'   Not as above...**2**
2   Frons with a black T-spot superiorly, abdomen chiefly dark brown or black......**3**
2'   Entirely or mostly reddish, virtually lacking conspicuous markings................**4**
3   Larger, abdomen (excl. apps.) male 50-57, female 54-64 mm; face bright green,
        superior transverse curved groove on the anterior surface of frons usually not
        black or dark brown..***viriditas***
3'   Smaller, abdomen (excl. apps.) male 44-46, female 47-48 mm; face bright blue
        (male) or green (female), the black T-spot of the frons extending on to the
        anterior surface and filling the superior transverse curved groove thereon
        ....***adnexa***
4   Male with prominent, strongly projecting genital lobe (Fig. AE6), cerci begin to
        expand at about 0.2 length, with conspicuous long hairs along inner edge of the
        cerci; female with brown wing veins and reddish brown hind tibiae........***diapyra***
4'   Male without prominent genital lobe (Fig. AE7), cerci begin to expand at about
        0.1 length, without conspicious long hairs along the inner edge; Female with
        reddish basal wing veins and black hind tibiae...**5**

5   Male with costa brown beyond the nodus, dorsal process of the terminal segment
       of the penis flattened, with separated patches of black spines; female mostly
       green .... *amazonica*
5'   Male with costa red to the pterostigma, dorsal process of the terminal segment of
       the penis with patches of black spines vertically oriented, parallel and almost
       touching; female reddish .... *perrensi*

Compiled from: CALVERT (1903, 1956), GEIJSKES (1943), PAULSON (1994)

Gynacantha

These are mostly large, slender, dull green or brown dragonflies often active at dusk or
dawn (DUNKLE, 1989; WILLIAMS, 1937; WILLIAMSON, 1923a). However, some species,
e.g. *G. tibiata*, are more active during the day. The latter also differs in general
coloration, being one of the most brilliantly colored aeshnids of all. It is easily
recognized by the combination of blue-green eyes, green thorax and orange-yellow end
of abdomen.
Although adults can be seen flying at a wide variety of locations, most species are
essentially forest insects and seem to be most numerous at comparatively low elevations
(WILLIAMSON, 1923a).

1   One or more antenodal crossveins of the second series, and rarely of the first
       series, basal to the first thickened antenodal, present in forewings or hind wings
       or both **2**
1'   No antenodal crossveins of the first or second series, basal to the first thickened
       antenodal, present (rarely a single crossvein may be seen in one wing) **3**
2   Colored basal wing area reduced, occupying less than half the median space and
       bounded posteriorly by A; T-spot on frons well marked. Male with the ventral
       carina on segment 2 with few to no teeth anterior to the point of convergence and
       a few posterior to the same point and with seven or eight denticles on the auricle
       .... *gracilis*
2'   Color of basal wing area extensive, occupying all or nearly all the median space
       and extending posteriorly beyond A, especially in the hind wing; frons above
       black or obscure, no evident T-spot. Male with the ventral carina on segment 2
       with a row of ten or more teeth anterior to the point of convergence and none
       posterior to this point, and with only three or fourth denticles on the auricle
       .... *membranalis*
3   Medium to large (abd. 43-62, hind wing 42-57 mm), dull colored, brownish
       insects; sides of thorax brown with at least four usually distinctly defined brown
       to black spots or areas as follows: one surrounding the metastigma, a spot above
       the metastigma, a spot at the upper end of the second lateral suture and a spot or
       stripe posteriorly on the latero-ventral carina; male with ventral carina on
       segment 2, posterior to the point of convergence, never concave, meeting the
       lateral carina in a more or less rounded angle **4**

3'   Small to medium insects (abd. 32-54, hind wing 32-54 mm), dull to brilliant coloration; ventral carina on segment 2, posterior to the point of convergence, concave, meeting the lateral carina at an acute angle **5**

4   A wide black stripe covering the posterior third or more of the metepimeron; wings hyaline or with the merest trace of color near the coastal margin; male with the abdomen greatly constricted at segment 3 when seen dorsally; female with the lateral carina on segment 2 distinctly black .... ***mexicana***

4'   Not as above; male with wings hyaline or uniformly brownish tinged; male with the abdominal segment 3 slightly or not constricted when seen in dorsal view; female with the lateral carina not distinctly black .... ***nervosa***

5   Legs pale colored, yellowish or reddish, four posterior femora with no distinct black **6**

5'   Legs more or less black, at least the apices of the four posterior femora black ... **8**

6   No distinct brown stripe on the first lateral suture; anal loop normally separated by two or three rows of cells from the posterior wing margin **7**

6'   A distinct brown stripe on the first lateral suture; anal loop separated by a single row of cells from the posterior wing margin .... ***caudata***

7   Hind wing less than 45 mm; auricles of male in lateral view not extending caudad to the level of the transverse carina at mid-height; in the female the lateral and ventral carinae on segment 2 subparallel .... ***laticeps***

7'   Hind wing 45 mm or more; auricles of the male very large, expanded, in lateral view extending caudad far beyond the level of the transverse carina at mid-height; in the female the lateral and ventral carinae on segment 2 diverging anteriorly .... ***auricularis***

8   Pterostigma light yellowish brown; femora of middle and hind pair of legs largely brown superiorly with less than the apical third black .... ***helenga***

8'   Pterostigma dark brown or reddish brown; femora of middle and hind pair of legs nearly all black superiorly; **9**

9   Pterostigma dark brown; males with abdominal segments 8-10 black .... ***jessei***

9'   Pterostigma reddish brown; males with abdominal segments 8 and 9 orange, passing into a light yellow on 10 .... ***tibiata***

Derived from: WILLIAMSON (1923a, 1930)

Neuraschna

Neuraschna maya

Large sized dragonfly (abdomen 62 to 69 mm, excl. appds.). Thorax predominantly brown with pale markings as follows: on each lateral side of dorsum a pale ovoid spot, lateral sides with two complete pale stripes parallel to the sutures, one on the mesepimeron, the other on the metepimeron. Abdomen dark brown, that of male swollen at base and constricted at segment 3, then almost parallel to base of segment 7, then gradually widening to apex of segment 8. Females with abdomen not constricted at segment 3 and then narrowing to apex of segment 7. Adults seem to prefer swampy wooded areas, where they hunt in the dusk along forest edges and along river banks low over bushes (BELLE, 1989a). Larvae have been collected in temporary brooklets in deep primary forest, where adults seldomly could be seen (CARVALHO, 1989).

Oplonaeschna

Oplonaeschna armata

Brownish dragonflies with rather short thick abdomen. Face pale greenish with yellow labrum. Two narrow pale green antehumeral stripes - one transverse in front of the antealar sinus, the other longitudinal-vertical; or these two united into one stripe, which thereby takes the form of an inverted L. Sides of thorax with two bluish stripes becoming yellowish below where they are bordered by black, with a general background of brown. Wings hyaline with yellow or brown pterostigma and rather conspicuous black spots upon and above the base of the wing roots. Abdominal segments 2-8 on dorsum with two pairs of yellow spots, the hinder pair well separated. Abdomen 44-53 mm in males, 46 mm in females (CALVERT, 1908; NEEDHAM, 1929).
Males fly for long distances low over the water of small rocky mountain streams (DUNKLE, in litt.).

Remartinia (formerly *Coryphaeschna* in part)

Large sized dragonflies similar to *Coryphaeschna*, from which genus both species were removed by CARVALHO (1992). They are separated predominantly by characters of wing venation and male secondary genitalia.

1   Male superior appendages excised on the inner margin; fork of IR3 being symmetrical_____***luteipennis***
1'   Male superior appendages not excised; fork of IR3 not symmetrical, the upper branch being a continuation of the vein in direction and the lower branch springs from this_____***secreta***

Derived from: CALVERT (1956), CARVALHO (1992)

Staurophlebia

Staurophlebia reticulata

Large sized dragonflies (up to 95 mm total length). Eyes anf thorax green, abdomen light brown except from first to basal half of third segment, and dorsum of tenth segment being yellowish-green.
This species can be found along creeks, banks of rivers, and bush paths. It is active during the day and flies mostly at a level from 5-10 m high (GEIJSKES, 1959).

Triacanthagyna

Fairly large sized dragonflies. Eyes blue to green, thorax green with brown or black markings. Abdomen reddish brown to almost black, with light greenish markings. As in *Gynacantha*, most species are mainly crepuscular, flying during the last hours of the day (WILLIAMSON, 1923a). Inside dark forests they may be found active also during the day.

1	Legs more or less dark (only mature individuals); thorax with definite dark markings; abdomen dark; anterior edge of frons, seen from above more or less angled; male with abdomen constricted at segment 3 .... **2**
1'	Legs entirely pale; thorax without definite dark markings; abdomen pale; anterior edge of frons seen from above convex; male with abdomen not constricted at segment 3 .... ***septima***
2	Anterior row of cells in anal loop usually consisting of two cells; superior appendage of male with the sides of the blade beyond the narrowed base parallel; female with abdomen very slightly constricted, anal appendages about as long as the last three segments .... ***ditzleri***
2	Anterior row of cells in anal loop usually consisting of three cells **3**
3	Second and third femora dissimilar in color; superior appendage with the sides of the blade beyond the base slightly converging posteriorly as seen in supero-internal view; female with abdomen not constricted, and the appendages slightly shorter than the last three segments .... ***caribbea***
3'	Second and third femora similar in color; superior appendage with the sides of the blade beyond the base parallel when see in supero-internal view; female with the abdomen constricted at segment 3 and the appendages as long as the last three and one-half segments **4**
4	Hamular process (see Fig. AE9) less than 0.6 mm long; margins of genital fossa without spines; females as in 3' .... ***trifida***
4'	Hamular process about 0.7 mm long; posterior border of genital fossa with a large patch of black spines or teeth; female unknown .... ***satyrus***

After: WILLIAMSON (1923a)

74

Gomphidae

Plate XIII, p. 75

Small to large sized dragonflies (25-85 mm body length) with male abdomen more or less expanded at segments 8 and 9. They are generally brightly patterned with black, green, and yellow stripes and spots. Most species fly during the early rainy season, the time when most matings occur (DONNELLY, 1992).

Key to the genera of Gomphidae

1	Occipital crest not ridged	**2**
1'	Occipital crest ridged (Figs. G1, G2)	**5**
2	Wings with 2-3 cubito-anal crossveins	***Epigomphus*** (p. 78)
2'	Wings with one cubito-anal crossvein	**3**
3	Basal subcostal crossvein (bsc) present (Fig. G3); Male face largely bluish green, pterothorax green with dark brown stripes, abdomen widest on segments 1 and 2 ....***Agriogomphus*** (p. 76)	
3'	Basal subcostal vein absent	**4**
4	Male anal triangle in hind wing absent, vulvar lamina with two very slender cross-laid, tapering branches extending beyond apex of ninth sternum (Fig. G4); females with a pair of long, posteriorly directed horns on the rear margin of the occiput ....***Archaeogomphus*** (p. 77)	
4'	Male anal triangle in hind wing present, vulvar lamina as long as segment 9, and divided a little more than halfway to base, divisions stout (Fig. G5); face black with bluish green markings, pterothorax blackish brown with green stripes, legs brown ....***Perigomphus*** (p. 84)	
5	Males	**6**
5'	Females	**11**
6	Anal triangle in hind wing not extending backward to anal angle (Fig. G6)	**7**
6'	Anal triangle in hind wing extending backward to anal triangle	**9**
7	Posterior genital hamule (Fig. G7) broadly expanded at base, with an anterior row of denticles or a chitinous ridge ....***Progomphus*** (p. 88)	
7'	Posterior genitale hamule of different shape	**8**
8	First anal interspace (x) in hindwing greater than second anal interspace (y) (Fig. G8) ....***Erpetogomphus*** (p. 82)	
8'	First anal interspace in hindwing smaller than second anal interspace; abdomen predominantly dark brown, becoming blackish on apical segments ....***Desmogomphus*** (p. 77)	
9	Penial peduncle reduced to a transverse lamella ....***Aphylla*** (p. 76)	
9'	Penial peduncle ('vesicle') with two more or less parallel plates	**10**
10	Flagella of penis strongly curled, finely serrulated along outer margin (Fig. G9) ....***Phyllocycla*** (p. 85)	
10'	Flagella of penis slightly bent, not serrulated (Fig. G10) ....***Phyllogomphoides*** (p. 86)	
11	Supratriangles usually with crossveins	**14**
11'	Supratriangles usually uncrossed	**12**

Plate XIII

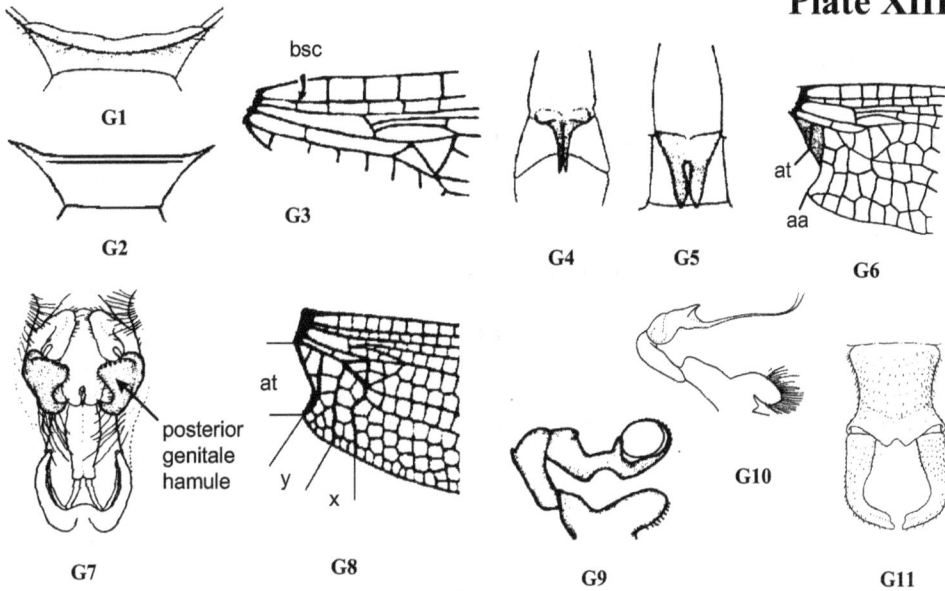

Fig. G1-G11:
G1: Occipital plate of *Progomphus gracilis* (female), dorsal view, showing occipital ridge;
G2: Occipital plate of *Aphylla obscura* (female), frontal view, showing occipital ridge; **G3**: Base of forewing of *Agriogomphus tumens* showing basal subcostal crossvein (bsc); **G4**: Vulvar lamina of *Archaeogomphus furcatus*; **G5**: Vulvar lamina of *Perigomphus pallidistylus*; **G6**: Base of hind wing of *Progomphus maculatus* showing male anal triangle (at) not extending backwards to anal angle (aa); **G7**: Second abdominal segment of *Progomphus gracilis*, ventral view, showing posterior genitale hamule with broadly expanded base; **G8**: Base of hindwing of *Phyllogomphoides brunneus* (male) showing first anal interspace (x), second anal interspace and anal triangle (at); **G9**: Penis of *Phyllocycla volsella* showing serrulated flagella; **G10**. Penis of *Phyllogomphoides semicircularis* showing two long flagella, not serrulated; **G11**: Abdominal segment 10 of *Aphylla obscura* showing dorsal apical rim

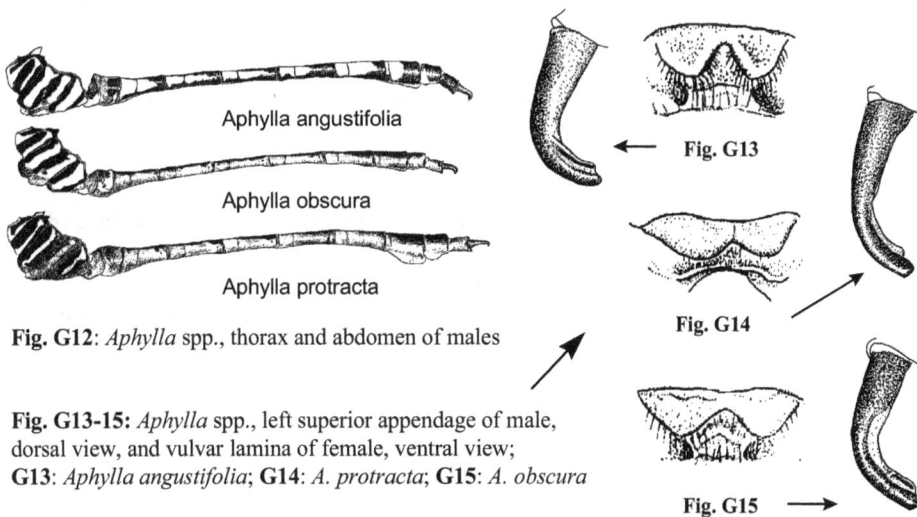

Fig. G12: *Aphylla* spp., thorax and abdomen of males

Fig. G13-15: *Aphylla* spp., left superior appendage of male, dorsal view, and vulvar lamina of female, ventral view;
G13: *Aphylla angustifolia*; **G14**: *A. protracta*; **G15**: *A. obscura*

12   Vulvar lamina extending well beyond middle of ninth sternum
        ....***Desmogomphus*** (p. 77)
12'   Vulvar lamina not extending beyond middle of ninth sternum.......................**13**
13   Distinct auricle rudiments present (visible on side view of second abdominal
        segment)..***Erpetogomphus*** (p. 82)
13'   Auricles completely absent...***Progomphus*** (p. 88)
14   Abdominal segment 10 without a dorso-apical rim; males with inferior
        abdominal appendage present...***Phyllogomphoides*** (p. 86)
14'   Abdominal segment 10 with dorso-apical rim (like Fig. G11 but narrower),
        males without inferior abdominal appendage...**15**
15   Distal spines of outer row on third femur as long as, or shorter than, one-sixth
        femur diameter, distal abdominal sterna orange or reddish brown
        ....***Aphylla*** (p. 76)
15'   Distal spines of outer row on third femur as long as, or longer than, one-fourth
        femur diameter, distal abdominal sterna not orange or reddish brown
        ....***Phyllocycla*** (p. 85)

After: BELLE (1988)

Agriogomphus

Agriogomphus tumens

Species of this genus are characterized by a pair of submedian spines near the hind margin of the occiput (better developed in females). Males of *A. tumens* have a largely bluish green face, their pterothorax being green with dark brown stripes, femora pale, becoming dark brown on distal ends, and the abdomen slender, widest on segments 1 and 2 (BELLE & QUINTERO, 1992). Adults usually perch quietly in shaded undergrowth near forested streams (DUNKLE, in litt.). The larvae have been found under decayed leaves deposited near mud banks of rain forest streams where the water flow is slow (NOVELO GUTIÉRREZ, 1989).

Aphylla
Plate XIII, p. 75

Males of this genus differ from all other dragonflies except *Phyllocycla* by the lack of the inferior appendage. They have the lower, apical angle of their 10th abdominal segment generally prolonged backward to a point, while the female's 7th-9th abdominal segments are noticably widened (BELLE & QUINTERO, 1992). They can be found perching on tips of sticks at lentic habitats (DUNKLE, in litt.).

1   Males..**2**
1'   Females..**4**
2   Greatest width of laterale foliate margin of abdominal segment 8 ≥ 0.80 mm (see
        Fig. G12), superior appendage as in Fig. G14.................................... ***protracta***

2'   Greatest width of laterale foliate margin of abdominal segment 8 ≤ 0.60 mm **3**

3   Apical mesal margin of superior appendage in dorsal view forming an elevated
        ridge that extends anteriorly over apical third of appendage (Fig. G13). Anterior
        half of dorsum of abdominal segment 7 pale yellow (Fig. G12); greatest width of
        foliate margin of abdominal segment 8 0.30-0.56 mm; posterior margin of abd.
        segment 10 dorsally nearly entire or emarginate with a shallow V-shaped notch
        ....***angustifolia***

3'   Apical mesal margin of superior appendage smoothly curved, not overlying
        apical third of appendage (Fig. G15). Anterior half of dorsum of abd. segment 7
        not pale, but the same color as posterior half (Fig. G12); greatest width of foliate
        margin of abd. segment 8 0.20-0.32 mm; notch of abd. segment 10 deep, U- or
        V-shaped..***obscura***

4   Greatest width of lateral foliate margin of abd. segment 8 ≥ 0.50 mm ...***protracta***

4'   Greatest width of lateral foliate margin of abd. segment 8 ≤ 0.30 mm **5**

5   Dorsum of anterior half of abd. segment 7 pale yellow, contrasting with dark
        brown of posterior half of that segment; width of lateral foliate margin of abd.
        segment 8 0.20-0.30 mm..***angustifolia***

5'   Dorsum of segment 7 concolorous, with a thin, pale middorsal line; width of
        lateral foliate margin of segment 8 usually narrower than for A. angustifolia,
        0.16-0.22 mm..***obscura***

After: GARRISON (1986)

Archaeogomphus

Archaeogomphus furcatus

Males of this genus are unique in the form of the 10th abdominal segment. „This has the
dorsum armed basally on either side with a strong interno-posteriorly directed hook and
produced apically in a long snout-like projection which is more than twice as long as the
rudimentary, flap-like, superior appendages;..." (WILLIAMSON, 1919a). Males of *A.
furcatus* having the ventral margin of the penial peduncle, in posterior view, deeply
excised V-shaped. Females are easily distinguished from other species by a pair of long,
posteriorly directed spines on the rear margin of the occiput.
Adults are found perching on tips of twigs in forests, or within streamside bushes
(DUNKLE, in litt.).

Desmogomphus

Desmogomphus paucinervis

Pterothorax brown with greenish-yellow stripes; legs brown; abdomen predominantly
dark brown, becoming blackish on apical segments (BELLE, 1972).

Epigomphus

Plates XIV (p. 79) & XV (p. 81)

Medium sized dragonflies occuring mostly on streams in forested habitats, although adults may be found also in more open areas along forest edges. Males have their abdomen generally widest at segment 10 (CALVERT, 1920). Many species look alike except for distinctive male appendages and female head structures.

Males

1   One antehumeral stripe and a more posterior, superior, antehumeral spot representing the upper end of the second antehumeral stripe (Fig. G32)............**2**

1' Two narrow antehumeral stripes, the posterior one very close to the humeral suture (Fig. G43)..**11**

2   Abdominal segment 10 with a dorsal tubercle, spinulose at tip, somewhat elongated transversely and divided by a shallow median emargination into two halves. Branches of superior appendages in dorsal view slightly divergent, the apex being truncated slightly obliquely; in profile apical fourth rather abruptly contracted both above and below. Inferior appendage longer than the superiors, directed upward, bifid in its apical three-fifth, bearing a strong, carinate, superior tooth at slightly more than half its length...*camelus*

2' Abdominal segment 10 without a dorsal tubercle, anal appendages not as above
    ....**3**

3   Apex of superior appendage in lateral view more or less straight or rounded, not forming distinct acute tips (Fig G27)..**6**

3' Superior appendage distinctly bifurcate apically (Fig. G16)............................**4**

4   Apex of superior appendage emarginated almost in a semicircle, forming two rather acute tips, only the upper one clearly visible in dorsal view, superior appendage in dorsal view with its inner edge being waved concavely, its outer edge convex; inferior appendage three-fourths as long as superiors, widely bifid in its apical half, each branch with an acute superior basal tooth near outer margin. (Figs. G16-18)..*quadracies*

4' Not as above...**5**

5   Inferior appendages distinctively longer than superiors (Fig. G19), branches divergent at base, parallel in apical half, at apices a pair of short upturned teeth (Fig. G20), thorax as in Fig G21..*echeverrii*

5' Inferior appendages slightly longer than superiors or almost equal in length (Fig. G23), branches divergent over the whole length, each one bearing a pair of short apical teeth, outer apical projection of superior appendage curls ventrally (Figs. G24 & 25)...*subsimilis*

6   Inferior appendages close to base only slightly divergent, posterior margin more or less V-shaped (slightly acute)...**7**

6' Inferior appendages close to base strongly divergent, posterior margin widely U-shaped, each branch slightly bifid apically, on the superior surface at the base with a slender forwardly curved spine. Superior appendage with apex rounded, the inner surface of the branch with a tubercle or small spine shortly before its middle (Fig. G26)...*verticicornis*

Plate XIV

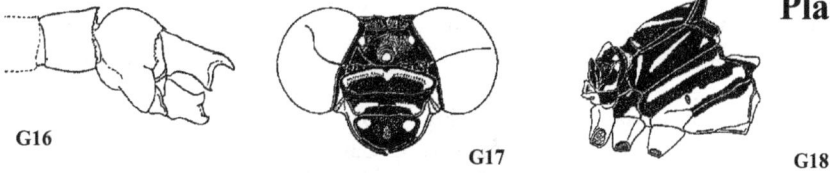

Figs. G16-18: *Epigomphus quadracies,*
G16: apex of abdomen, lateral view; **G17**: head, frontal view; **G18**: thorax, lateral view

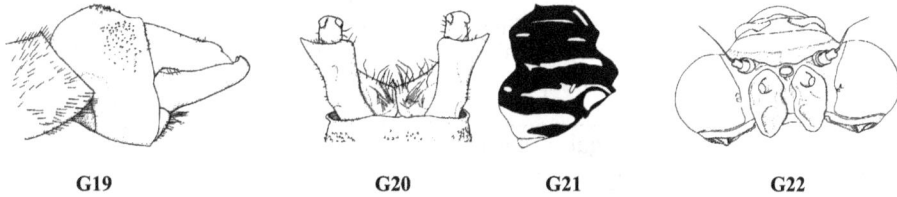

Figs. G19-22: *E. echeverrii,* **G19**: appendages, lateral view; **G20**: appds., dorsal view;
G21: thoracic color pattern; **G22**: head, dorsal view

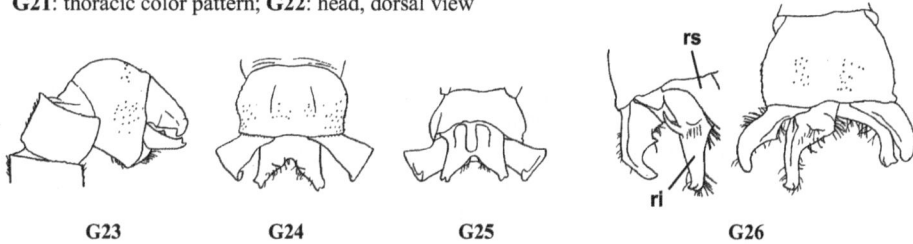

Figs. G23-25: *E. subsimilis,* **G23**: apex of abdomen,
lateral view; **G24**: dorsal view of appendages;
G25: ventral view of appendages

Fig. **G26**: *E. verticicornis,*
abdominal appds., dorso-lateral
view (left) and dorsal view (right),
rs - right superior appd.,
ri - right inferior appd.

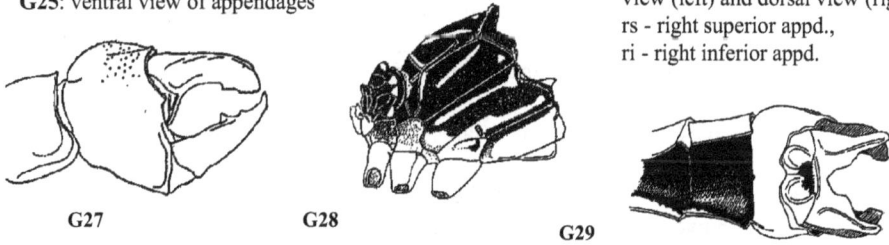

Figs. G27-29: *E. subquadrices*
G27: apex of abdomen, lateral view; **G28**: thorax, lateral view; **G29**: apex of abdomen, ventral view

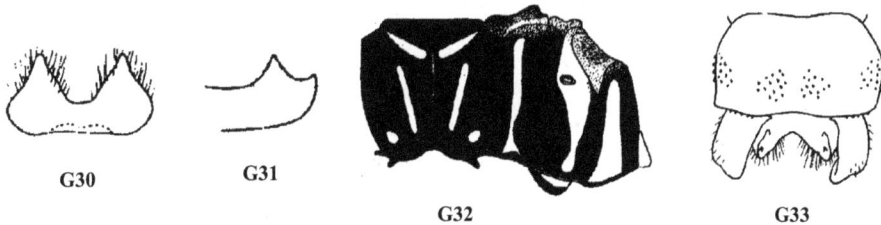

Figs. G30-33: *E. compactus*
G30: inferior appd., posterior view; **G31**: right branch of inferior appd., left lateral view;
G32: thoracic color pattern; **G33**: apex of abdomen, dorsal view

7   Inferior appendages deeply bifurcate with stepped mesal edges of the two branches (Fig. G29), dorsal surface of the inferior appendage with three teeth beyond its middle (Fig. G27), thorax as in Fig. G28._____*subquadrices*

7'   Not as above._____**8**

8   Short and compact caudal appendages (Fig. G33), the stocky branches of the inferior appendage are vertically upraised (Fig. G30) and viewed laterally they are only discernible to a small extent (Fig. G31); thoracic color pattern as in Fig. G32_____*compactus*

8'   Not as above_____**9**

9   Superior appendages with apex obliqueley truncated dorsad and caudad. Inferior appendages distinctly shorter than the superiors, bifid in its apical half, each branch at its apex with 4-5 superior denticles in a row parallel to its inner margin._____*tumefactus*

9'   Superior appendage rectangular, dorsal edge curving ventrally in apical half; inferior appendages as long as or longer than superiors (Fig. G36)_____**10**

10   Pale antehumeral stripe separated from the pale mesothoracic transverse stripe (Fig. G34); superior appendage with seven short teeth on ventro-apical margin. Inferior appendages deeply bifurcate, apices in dorsal view deeply hooked and curving inwards, each branch bearing a small median spine in a shallow depression (Fig. G35)_____*houghtoni*

10'   Pale antehumeral stripe well connected with the mesothoracic transverse stripe (Fig. G37); superior appendages vertically flattened (Fig. G38); rear of head bulged out near the upper edge of each compound eye._____*jannyae*

11   Inferior appendages in dorsal view V-shaped (Figs G40, G41)_____**12**

11'   Inferior appendages in dorsal view widely U-shaped (Fig. G39)_____*corniculatus*

12   Apex of inferior appendage not bifid._____**13**

12'   Apex of inferior appendage bifid, terminating in two distinct small teeth (Fig. G44), posterior margin of inferior appendages deeply V-shaped (Fig. G45); thorax as in Fig. G43)_____**14**

13   Inferior appendages deeply V-shaped, each branch terminating with prominent rounded knobs (Fig. G41)._____*westfalli*

13'   Inferior appendages widely V-shaped, apices of branches without knobs (Fig. G40)_____*armatus*

14   Inferior appendage a little longer than the lower edge of superior appendage, in dorsal view with two teeth near the posterior margin (Fig. G45)_____*clavatus*

14'   Inferior appendages without any superior teeth, as long as superior appendages; superior appendage with its apex obliquely truncated caudad and mesad, in profile view curved downward sharply in its apical half (Fig. G42)____*subobtusus*

Females

1   One pale green or yellow antehumeral stripe and a more posterior pale green or yellow antehumeral spot, the latter representing the upper end of the second antehumeral stripe_____**2**

1'   Two narrow pale green or yellow antehumeral stripes, the posterior one close to the humeral suture._____**5**

Plate XV

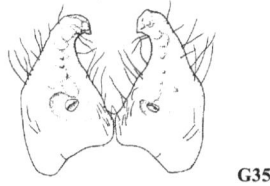

G34   G35

Figs. G34 & 35: *Epigomphus houghtoni*
G34: thoracic color pattern;
G35: inferior appendages, dorsal view

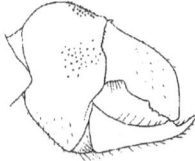

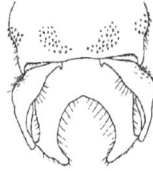

Figs. G36-38: *E. jannyae*
G36: apex of abdomen,
lateral view; **G37**: thoracic
color pattern; **G38**: abd.
appendages, dorsal view

G36   G37   G38

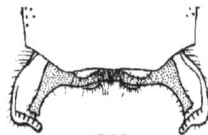
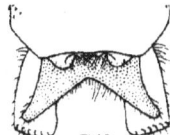
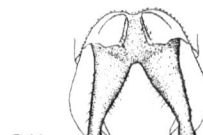
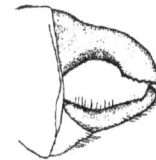

G39   G40   G41

Figs. G39-41: *Epigomphus* spp., apex of male abdomen, ventral view;
G39: *E. corniculatus*, **G40**: *E. armatus*, **G41**: *E. westfalli*

Fig. G42: *E. suboptusus*,
male abd. appendages,
lateral view

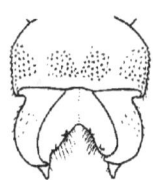
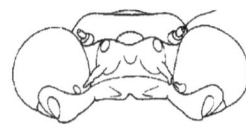

G43   G44   G45

Fig. G46: *E. corniculatus*,
head of female, dorsal view

Fig. G43-45: *E. clavatus*,
G43: thoracic color pattern, **G44**: abdominal appds., lateral view, **G45**: abdominal appds., dorsal view

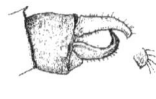

G47	G48	G49	G50	G51	G52
E. constrictor	*E. sabaleticus*	*E. tristani*	*E. ophibolus*	*E. bothrops*	*E. viperinus*

Figs. G47-57:
Erpetogomphus,
lateral (above) and dorsal
(below) view of abdominal
appendages of males

G53	G54	G55	G56	G57
E. schausi	*E. eutainia*	*E. leptophis*	*E. elaphe*	*E. elaps*

2   Behind each lateral ocellus no tubercle projecting markedly above the level of
        the ocellus itself._____**3**

2'   Behind each lateral ocellus a stout tubercle projecting much above the level of
        the ocellus itself_____*verticicornis*

2''   Around each lateral ocellus a large flattened lobe which projects considerably
        from the rear of the head (Fig. G22)._____*echeverrii*

3   Vertex without five longitudinal grooves, occiput without a strong postero-
        superior tubercle at each lateral extremity._____**4**

3'   Vertex with five longitudinal grooves, one median, two lateral ocellar and two
        paraocular; occiput with a strong posterior or postero-superior tubercle, or
        rounded horn, at each lateral extremity._____*tumefactus*

4   In dorsal view each lateral ocellus subequally distant from the mid-dorsal line of
        the head and from the adjacent eye-margin._____*subsimilis*

4'   In dorsal view each lateral ocellus two to four times as far from the mid-dorsal
        line of the head as from the adjacent eye-margin (Fig. G17)_____*quadracies*

5   Behind each lateral ocellus no tubercle projecting dorsad markedly above the
        level of the ocellus itself._____*subobtusus & westfalli*

5'   Behind each lateral ocellus a stout tubercle which projects dorsad markedly
        above the level of the ocellus itself. Rear of the head with a much deeper pit each
        side than in the preceding species._____**6**

6   Posterior margin of occiput with a small but distinct median excision (Fig. G46)
        _____*corniculatus*

6'   Posterior margin of occiput medially slightly concave_____*armatus*

Compiled from: BELLE (1980, 1989b, 1993, 1994), BROOKS (1989), CALVERT (1920b),
DONNELLY (1986), KENNEDY (1946), RIS (1918)

Erpetogomphus

Plates XV (p. 81) & XVI (p. 87)

Adults of this genus „....are most commonly found near shores of streams and rivers, but
they may also be found in agricultural stubble or on tree branches near streams. Many
species are seldomly encountered..." (GARRISON, 1994).

Males

1   Superior appd. with a prominent superior tooth at about 0.75 of appendage
        length (Figs. G47-G50)_____**2**

1'   Superior appd. with dorsal surface convexly angulate, smoothly curved, straight,
        or with a concavity_____**5**

2   Inf. appd. smoothly curved, with tip pointing dorsally or posterodorsally (Fig.
        G50); a small sharp anteriorly directed superior tooth at middle of epiproct
        _____*ophibolus*

2'   Inf. appd. strongly curved so that distal 0.30 is parallel to basal 0.30 (Fig. G47);
        tip of epiproct pointing anteriorly, superior surface of epiproct with no tooth____**3**

3   Ventral margin of superior appd. smoothly concave (Fig. G47)_____*constrictor*
3'   Ventral margin with a large ventral tubercle at 0.5-0.75 length of superior appd. (Figs. G48, G49)_____**4**
4   In lateral view, superior tooth of superior appd. as long as rest of appd., so that appd. appears to end in two equal branches (Fig. G48)_____*sabaleticus*
4'   In lateral view, superior tooth of superior appd. less than half as long as remainder of appendage (Fig. G49)_____*tristani*
5   Basal part of superior appd. with no ventral carina, though a small tubercle may be present at basal 0.15 to 0.20 of appendage_____**6**
5'   Basal 0.25 to 0.30 of lower margin of superior appd. with a distinct longitudinal carina which may end in a small ventral tooth, appendages as in Fig. G54; with a dark stripe bordering posterior margin of metepimeron_____*eutainia*
6   Ventral margin of superior appd. straight or forming a gentle concave curve so that appendage appears linear (Figs. G55-G57)_____**7**
6'   Apical 0.30 of ventral margin of superior appd. curved downward (Figs. G51-G53)_____**9**
7   Thorax blue-green with full complement of dark thoracic stripes; face blue-green with extensive dark brown frontoclypeal stripe; appendages as in Fig. G55
       _____*leptophis*
7'   Thorax entirely yellow-green, or with only an antehumeral and faint indication of a dark middorsal and antehumeral stripe; face entirely pale_____**8**
8   Distal branch of anterior hamule thick, as high as gap separating the two branches; posterior hamules with no distal tooth; appendages as in Fig. G56
       _____*elaphe*
8'   Distal branch of anterior hamule narrow, not as high as gap separating the two branches; posterior hamules with a distal tooth; appendages as in Fig. G57 *elaps*
9   Thorax blue-green, with full complement of dark stripes, including second lateral stripe and stripe on posterior margin of metepimeron, with a well defined blue green spot on frons; appendages as in Fig. G53_____*schausi*
9'   Thorax blue-green or yellow-green; sides lacking complete dark second lateral stripe (except in some viperinus, incomplete in some bothrops) and without dark stripe on posterior margin of metepimeron; face predominantly pale, with dark areas limited to sutures_____**10**
10   Tip of inf. appd. in posterior view terminating in a blunt point; lateral and dorsal view of appendages as in Fig. G52_____*viperinus*
10'   Tip of inf. appd. in posterior view truncate or bidentate; in lateral view curved, extending 0.75 to 0.9 the length of cerci (Fig. G51); dark antehumeral and humeral stripes separate_____*bothrops*

Females

1   Medial area of occiput with a posteriorly pointed protuberance (Fig. G63)
       _____*leptophis*
1'   Medial area of occiput variously shaped, but never with a posteriorly directed protuberance_____**2**
2   Median ocellus posterior to lateral ocelli, the former lying within a deep longitudinal trough (Figs. G58, G59)_____**3**

2'   Median ocellus at level of or anterior to lateral ocelli, no longitudinal trough
        (Fig. G61)..**4**

3   Dorsal surface of vertex with sides of trough roughly parallel, or only slightly
        converging anteriorly; Northeastern Mexico south to Costa Rica (Fig. G58)
        ....*constrictor*

3'   Dorsal surface of vertex with sides of trough strongly converging anteriorly;
        Costa Rica to Western Panama (Fig. G59)..*tristani*

3''   As above; Central Panama, Colombia, Venezuela ....*sabaleticus*

4   Vertex with transverse ocellar ridge bilobed behind median ocellus, occiput
        wide, forming a full semicircle (Fig. G61)...*schausi*

4'   Vertex with transverse ocellar ridge entire, low and almost vestigial, or absent,
        its lateral ends forming oval tubercles posterior to lateral ocelli (Figs. G62, G64)
        ....**5**

5   Vulvar lamina followed on segment nine by distinct and prominent semicircular
        ridge, never with a posteriorly directed arm; vertex and occiput as in Fig. G62
        ....*eutainia*

5'   Vulvar lamina followed on segment nine by a Y-shaped ridge.......................**6**

6   Lobes of vulvar lamina separated by an almost U-shaped interval 3 to 4 times as
        wide as either lobe
                Mexico south to Guatemala; vertex and occiput as in Fig. G64......*elaps*
                Guatemala to Costa Rica...*elaphe*

6'   Lobes of vulvar lamina separated by a triangular or semicircular interval 0.5 to
        1.0 the width of each lobe..**7**

7   A deep pit at anterior margin of frons anterolateral to median ocellus (Fig. G67),
        cleft between vulvar lamina plates wide, forming an obtuse arc greater than
        100°; dark humeral and antehumeral stripes combined or nearly so; larger
        ....*viperinus*

7'   Anterior margin immediately anterior to median ocellus forming a narrow V-
        shaped trough with base of postfrons (Fig. G66); cleft between vulvar lamina
        plates narrow, forming a V-shaped notch of not more than 90°; dark humeral and
        antehumeral stripes separate; smaller.. *bothrops*

After: GARRISON (1994)

Perigomphus

Perigomphus pallidistylus

Face black with bluish green markings; pterothorax blackish brown with green stripes;
legs brown, except for the greenish inner side of fore femora. Abdomen dark brown on
basal segments, becoming black on apical segments including inferior anal appendages.
Superior appendages pale yellowish. Ventral tergal margins of segment 8 and 9
unexpanded (BELLE, 1972; BELLE & QUINTERO, 1992).

Phyllocycla
Plate XVI, p. 87

Males of this genus have foliaceous marginal expansions of the 8th abdominal tergum. Adults are found in forests around rivers, where they are difficult to detect (BELLE, 1988).

Males

1   Apical inferior angles of 10th abdominal segment, viewed ventrally, produced inward or inward and downward; upper margin of superior appendages in profile view gradually curved downward (Fig. G68)..**2**

1'   Not as above (Fig. G69); upper margin of superior appendage in profile view with a distinct tooth or strong angulation at the point where the appendage is down turned; Tip of superior appendage without internal hook; Labrum largely bluish-olive or with a pair of bluish-olive spots..*volsella*

2   Lateral dilatation of abdominal segment 9 sharply angled near base of segment
     ....*elongata*

2'   Lateral dilatation of abdominal segment 9 curved throughout............................**3**

3   Superior appendage, in dorsal view, nearly straight for proximal two-thirds; superior surface of frons largely leaden-grey to green...........................*breviphylla*

3'   Superior appendage, in dorsal view, evenly curved inward from base; superior surface of frons entirely brown..*speculatrix*

Females

1   Apical inferior angles of 10th abdominal segment, viewed ventrally, produced inward or inward and downward...**2**

1'   Not as above; Labrum largely bluish-olive or with a pair of bluish-olive spots
     ....*volsella*

2   Vulvar lamina two-fifth the length of ninth sternum; its posterior margin medially excised for more than three-fourth the length of vulvar lamina, the lobes slender...**3**

2'   Vulvar lamina one-third the length of ninth sternum or shorter; its posterior margin medially excised for less than two-thirds the length of vulvar lamina, the lobes strongly rounded..*speculatrix*

3   First pale antehumeral stripe for the greater part as wide as or narrower than the second pale antehumeral stripe immediately in front of the humeral suture; the green of metepimeron reaching to aslant hind border.................................*elongata*

3'   First pale antehumeral stripe for the greater part distinctly wider than the second pale antehumeral stripe immediately in front of the humeral suture; metepimeron with a green band on central part of this sclerite...........................*breviphylla*

After: BELLE (1988)

Phyllogomphoides

Plate XVI, p. 87

Large sized dragonflies with generally well-developed foliaceous expansions an 8th abdominal segment. They are mainly forest dwellers (BELLE, 1984c) inhabiting small shaded streams in forests (e.g., *P. insignatus*, *P. pugnifer*) but some species, like *P. suasus*, the largest species of the genus in Central America, fly in relatively open areas, such as along the banks of wide rivers (DONNELLY, 1979).

Males

1	Pale mesepimeral stripe present, sometimes reduced or narrow.	**2**
1'	Pale mesepimeral stripe undeveloped (Fig. G71)	*insignatus*
2	Second antehumeral stripe completely absent (Fig. G77) or reduced to a small spot on the upper end of mesepisternum	**3**
2'	Second antehumeral stripe present, sometimes very narrow and reduced in length	**5**
3	Second antehumeral stripe completely absent, pale metepisternal stripe well developed only in lower half (Fig. G77)	*litoralis*
3'	Small spot on the upper end of the mesepisternum, pale metepisternal stripe reduced	**4**
4	Inferior appendage more shallowly cleft, more widely V-shaped (slightly acute interval). Superior appendage viewed dorsally in its apical third distinctly more strongly angulate than in its lower part (Fig. G79)	*appendiculatus*
4'	Inferior appendages deeper V-shaped (distinctly acute intervall). Superior appendages viewed dorsally more or less uniformly angulate (Fig. G80)	*bifasciatus*
5	Inferior appendages widely cleft (almost rectangular) or U-shaped (Figs. G81, G82)	**6**
5'	Inferior appendages acutely V-shaped (Figs. G83, G84)	**7**
6	Inferior appendages U-shaped (Fig. G82). Superior appendages pale, with internal spine at more than 2/3 length, subapical shelf on inner surface produced into a rounded triangular process suggesting a second larger spine	*duodentatus*
6'	Inferior appendages widely cleft (Fig. G81). Superior appendages yellow, with internal spine at less than 2/3 length, subapical shelf on inner surface rounded	*pugnifer*
7	Anterior hamule rounded and flattened, narrowed apically and broadly excavated medially near tip	*suasillus*
7'	Anterior hamule flattened and rounded, inner margin neither notched nor excavated, the tip deflected mesally	*suasus*

Females (females of *pugnifer* and *litoralis* unknown!)

1	Pale mesepimeral stripe present, sometimes reduced or narrow.	**2**
1'	Pale mesepimeral stripe undeveloped	*insignatus*
2	Second antehumeral stripe present, sometimes very narrow and reduced in length	**3**

Plate XVI

G58
E. constrictor

G60
E. ophibolus

G59
E. tristani

G61
E. schausi

G62
E. eutainia

G63
E. leptophis

G64
E. elaps

G65
E. liopeltis

G66
E. bothrops

G67
E. viperinus

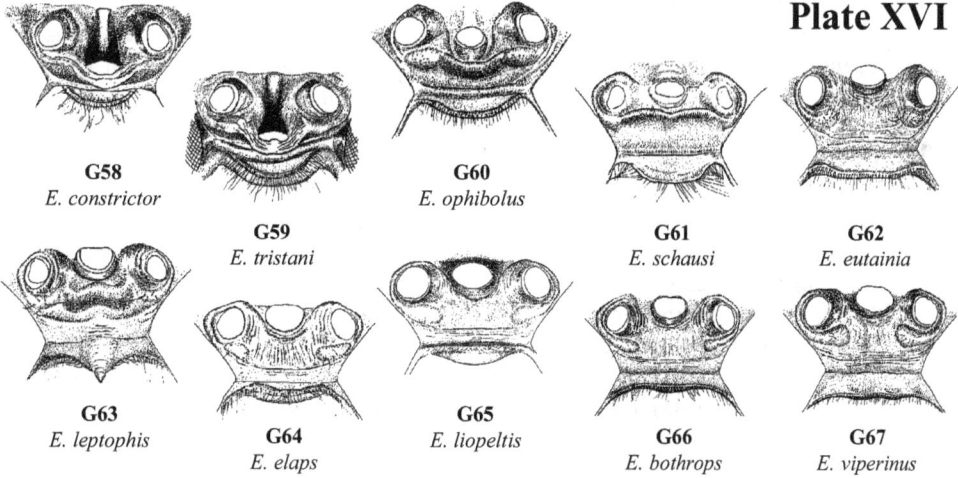

Figs. G58-67: *Erpetogomphus*, vertex & occiput of female, dorsal view

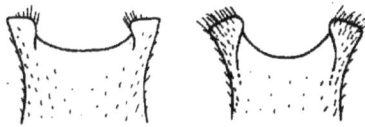

Figs. G68 & 69: *Phyllocycla*, ventral view of apex of 10th abdominal segment of males

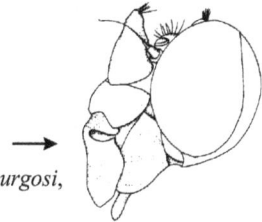

G68
P. elongata

G69
P. volsella

Fig. G70: *Phyllogomphoides burgosi*, head of female in lateral view

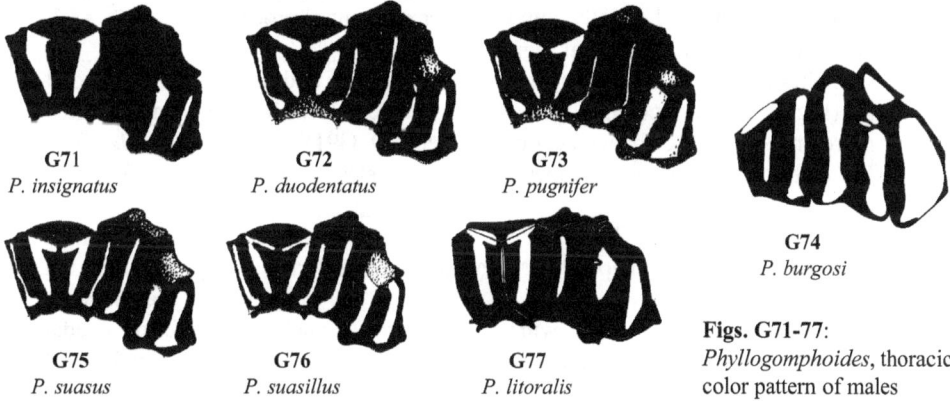

G71
P. insignatus

G72
P. duodentatus

G73
P. pugnifer

G74
P. burgosi

G75
P. suasus

G76
P. suasillus

G77
P. litoralis

Figs. G71-77: *Phyllogomphoides*, thoracic color pattern of males

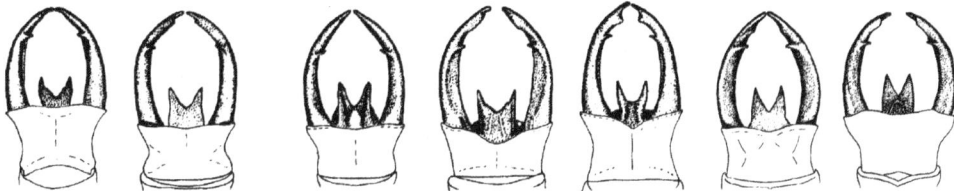

G78
P. insignatus

G79
P. appendiculatus

G80
P. bifasciatus

G81
P. pugnifer

G82
P. duodentatus

G83
P. suasillus

G84
P. suasus

Figs. G78-84: *Phyllogomphoides*, dorsal view of male abdominal appendages

2'   Second antehumeral stripe completely absent or reduced to a small spot on the upper end of mesepisternum..................................*appendiculatus & bifasciatus*

3   Frons with prominent lateral, dorsally projecting horn 2.8 mm in height (Fig. G70); vulvar scale broadly U-shaped, arms broad with small apical tubercle ....*burgosi*

3'   Not as above..**4**

4   First antehumeral stripe connected with the mesothoracic collar...................**5**

4'   First antehumeral stripe isolated from mesothoracic collar; vulvar lamina: median cleft sharp, sides produced ventrally, with lateroapical corner modified into a distinct laterally extended flat spine..............................*duodentatus*

5   Vulvar lamina tips rounded..*suasillus*

5'   Vulvar lamina tips truncate...*suasus*

Derived from: BELLE (1984a), BROOKS (1989), DONNELLY (1979)

Progomphus

Plate XVII, p. 91

1   Third tarsus about 2/3 the length of third tibia or shorter; dark midlateral and third lateral stripes of pterothorax narrow and not connected; caudal appendages of male as in Fig. G85...*pygmaeus*

1'   Third tarsus ¾ the length of third tibia, or longer..**2**

2   Abdominal segment 1 usually with a slender midventral process.....................**3**

2'   No such process on abdominal segment 1...**5**

3   Branches of inferior caudal appendage of male stout and inequilaterally bifid at tip (Fig. G90); postocellar ridges of vertex of female well developed and often more or less tubercle-shaped near eye border (Fig. G91)......................*clendoni*

3'   Branches of inferior caudal appendage of male slender; vertex of female not as above...**4**

4   Abdomen 32-39 mm, hind wing 27-30 mm; Branches of inferior caudal appendage of male with apical part beyond supero-external anteapical tooth long and strongly incurved (Fig. G86)..*mexicanus*

4'   Abdomen 40-43 mm, hind wing 32-33 mm; Branches of inferior caudal appendage of male very slender and widely separated, the apical part beyond supero-external anteapical tooth obsolete and no more than a low hump (Fig. G87)...*zonatus*

5   First pale antehumeral stripe more or less wedge-shaped, broad below, the upper portion tapering..**6**

5'   First pale antehumeral stripe not wedge-shaped; labrum with a symmetric pair of grey spots...*risi*

6   Posterior margin of occipital plate deeply concave, V-shaped (Fig. G92); pale antealar spot present; pterostigma longer than half the distance from nodus to pterostigma; anal appendages of male as in Fig. G89......................*longistigma*

6'   Posterior margin of occipital plate not markedly concave; no pale antealar spot;
         pterostigma shorter than half the distance from nodus to pterostigma; anal
         appendages of male as in Fig. G88⎯⎯⎯⎯⎯⎯⎯⎯⎯⎯⎯⎯⎯⎯⎯⎯⎯⎯⎯⎯⎯⎯***anomalus***

After: BELLE (1973)

Cordulegastridae

Cordulegaster godmani

Large sized dragonflies of approximately 75 mm in body length. Conspicuously colored
black with yellow stripes on thorax, black with a more or less complete yellow ring on
each of the abdominal segments. Females with an ovipositor projecting much beyond
apex of abdomen.
They occur in small streams in forested habitats at higher elevations, usually above
1700 m (ESQUIVEL, 1991).

Corduliidae

Only one genus, *Neocordulia*, is found in Central America.

Neocordulia

Plate XVII, p. 91

Medium sized dragonflies (35-40 mm abdomen length). Pterothorax with dark green metallic areas, with blue or purplish reflections. Abdomen mostly black or dark reddish brown dorsally and dorsolaterally, moderately widened beyond segment 6, not depressed dorsoventrally. Segment 8 of male without sternal protuberance (MAY, 1991).

1	Genital lobe extending beyond hamule for about 1/3 its own length (Fig. NC1); frons and vertex orange-brown to dark brown, not at all metallic; middorsal keel of abdominal segment 10 narrow..*campana*
1'	Genital lobe extending barely, if at all, beyond hamule; frons and vertex dark blue, metallic; middorsal keel of abdominal segment 10 broad.............................**2**
2	Hamules not especially robust, lateral ridges of dorsal and ventral branches narrow (Fig. NC2); cerci each of nearly uniform diameter for most of length, often with rather abrupt subapical swelling (Fig. NC4), shorter than 2.3 mm; metafemur usually shorter than 6.4 mm, hindwing usually shorter than 36 mm ....*batesi*
2'	Hamules robust, lateral ridges of dorsal and ventral branches thick (Fig. NC3); cerci in dorsal or mediodorsal view gradually increasing in diameter to widest point slightly before apex (Fig. NC5), longer than 2.5 mm; metafemur at least 6.4 mm long, hindwing at least 36 mm long..*griphus*

After: MAY (1991)

Plate XVII

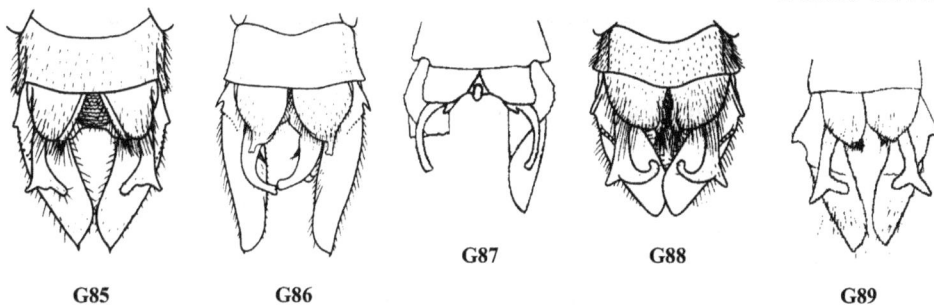

Figs. G85-89: *Progomphus*, apex of abdomen, ventral view **G85**: *P. pygmaeus*, **G86**: *P. mexicanus*, **G87**: *P. zonatus*, **G88**: *P. anomalus*, **G89**: *P. longistigma*

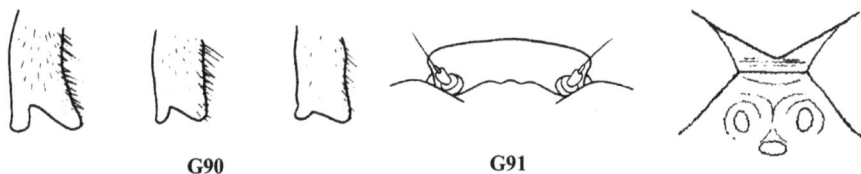

Figs. G90 & 91: *Progomphus clendoni*, **G90**: three variants of tip of male inferior abdominal appendage; **G91**: postocellar ridge of vertex of female

Fig. G92: *P. longistigma*, occipital plate, viewed from above

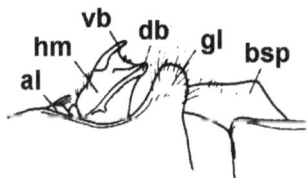

NC2: *N. batesi*

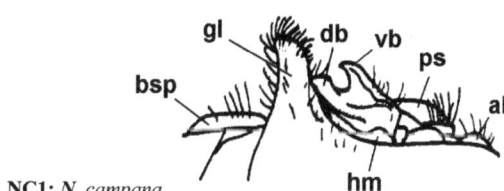

NC1: *N. campana*

Figs. NC1-3: *Neocordulia*, lateral view of male accessory genitalia

NC3: *N. griphus*

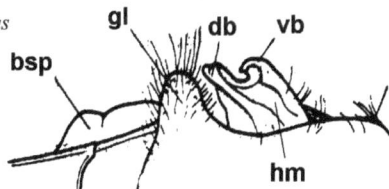

al - anterior lamina; bsp - basal segment of penis; db - dorsal branch of hamule; gl - genital lobe; hm - hamule; ps - penile shield; vb - ventral branch of hamule

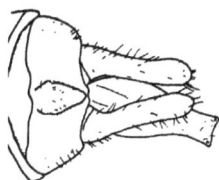

NC4: *N. batesi*

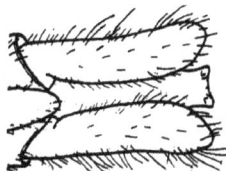

NC5: *N. griphus*

Figs. NC4 & 5: *Neocordulia*, dorsal view of male abdominal appendages

Libellulidae

Plate XVIII, p. 93

Key to the genera of Libellulidae

1   Costa of forewing with a concave indentation between base and nodus
                                                              Zenithoptera (p. 125)

1'   Costa of forewing without such an indentation **2**

2   Last antenodal in forewing complete (Fig. Li1) **3**

2'   Last antenodal in forewing incomplete (Fig. Li2) **30**

3   One bridge cross-vein **4**

3'   Two or more bridge cross-veins **20**

4   Arculus in hind wing between first and second antenodals **5**

4'   Arculus in hind wing opposite or distad of second antenodal **18**

5   R3 undulate (Fig. Li11) **6**

5'   R3 not undulate (Fig. Li12) **8**

6   Sectors of arculus in forewing stalked (Fig. Li4), 3 (rarely 4) cell rows in discoidal field (df) of forewing, the field parallel sided or slightly narrowed distally, 1 (rarely 2) cell rows between MA and Mspl in forewing **7**

6'   Sectors of arculus in forewing not stalked (Fig. Li3), 4 cell rows in df of forewing, the field widened distally, 2 cell rows between MA and Mspl in forewing .... ***Libellula*** (p. 110)

7   Hamules 2-parted (Fig. Li20); sides of abd. segment 8 in female expanded (Fig. Li23) .... ***Orthemis*** (p. 117)

7'   Hamules not 2-parted; sides of 8 in female not expanded .... ***Dythemis*** (p. 103)

8   Usually no crossveins behind pterostigma, if there is a crossvein it is behind the distal fourth of the pterostigma; a very long cell behind pterostigma (Fig. Li102)
                                                  Pachydiplax (p. 119)

8'   One or more crossveins behind pterostigma, usually at least with one crossvein behind proximal half of pterostigma; cells behind pterostigma usually of normal length **9**

9   CuP in hind wing arising at anal angle of triangle (Fig. Li5) **10**

9'   CuP in hind wing separated from anal angle of triangle (Figs. Li6) **16**

10   Triangle in forewing free **11**

10'   Triangle in forewing crossed **14**

11   Posterior lobe of prothorax narrowed at base (Figs. Li13), and usually erect and bearing a fringe of long hairs **13**

11'   Posterior lobe of prothorax not narrowed at base (Figs. Li14), and usually not erect or bearing a fringe of long hairs **12**

12   Mspl in forewing very distinct (Fig. Li101); 5-6 rows of cells in anal field of hind wing; posterior lobe of prothorax small, its distal margin weakly arched
                                                  Macrodiplax (p. 111)

12'   Mspl in forewing usually indistinct; 2-4 (rarely more) rows of cells in anal field of hind wing; posterior lobe of prothorax larger, rounded or quadrate
                                                  Erythrodiplax (p. 106)

13   Aspl in hind wing bent at level of heel (as in Figs. Li 7, 8) **91**

Plate XVIII

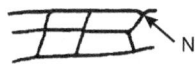

Fig. Li1: Last antenodal in front wing complete

Fig. Li2: Last antenodal in front wing incomplete

Fig. Li3: Sectors of arculus in front wing not stalked

Fig. Li4: Sectors of arculus in front wing stalked

Li5: CuP arising at anal angle of triangle

Li6: CuP separated from anal angle of triangle

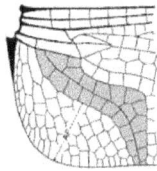

Fig. Li7: Anal loop open posteriorly and extending to wing margin

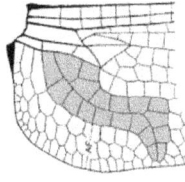

Fig. Li8: Anal loop closed posterior and not extending to wing margin

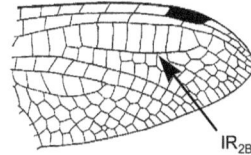

Fig. Li9: IR_{2B} well developed

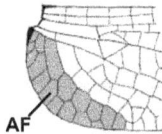

Fig. Li10: Anal field with two rows of cells

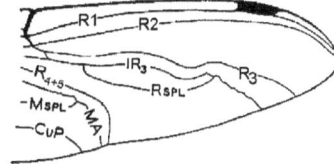

Fig. Li11: R3 very undulate (*Pantala*)

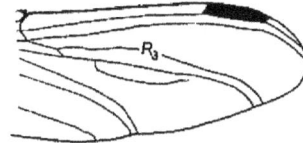

Fig. Li12: R3 not undulate (*Pachydiplax*)

Fig. Li13: Posterior lobe of prothorax narrowed at base (dorso-anterior view)

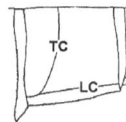

Fig. Li14: Posterior lobe of prothorax not narrowed at base (dorso-anterior view)

Fig. Li15: Lateral view of 2nd abdominal segment, showing lateral (LC) and transverse (TC) carina

Fig. Li16: Tooth on tarsal claw longer or nearly as long as tip of claw

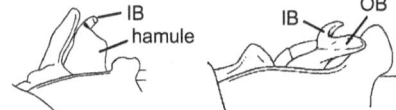

Fig. Li17: Tooth normal

Fig. Li18: Tooth small and near tip of claw

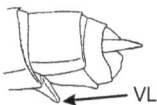

Fig. Li19 & 20: Accessory genitalia of male, lateral view; IB - inner branch, OB - outer branch

Fig. Li21: Vulvar lamina (VL) scoop shaped and projecting ventrad

Fig. Li22: Vulvar lamina (VL) in ventral view small and bilobed

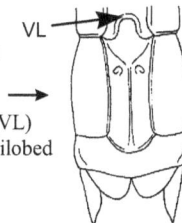

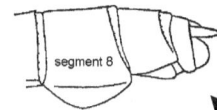

Fig. Li23: Sides of segment 8 of female expanded

13'	Aspl in hind wing straight or nearly so	**73**
14	Posterior lobe of prothorax not narrowed at base (Figs. Li14), and usually not erect or bearing a fringe of long hairs	**15**
14'	Posterior lobe of prothorax narrowed at base (Figs. Li13), and usually erect and bearing a fringe of long hairs	**25**
15	Two to three cell rows in discoidal field of forewing; ends of pterostigma parallel; a distinct subtriangle in forewing of 1-4 cells; hind wing not noticeably broadened at base	***Erythrodiplax*** (p. 106)
15'	Four cell rows in discoidal field of forewing; ends of pterostigma not parallel; subtriangle in forewing not distinct, several cells in this area of wing; hind wing noticeably broadened at base (Figs. Li104 & 105)	***Tramea*** (p. 124)
16	Posterior lobe of prothorax narrowed at base (Figs. Li13), and usually erect and bearing a fringe of long hairs	**17**
16'	Posterior lobe of prothorax not narrowed at base (Figs. Li14), and usually not erect or bearing a fringe of long hairs	**18**
17	Triangle in forewing free	**77**
17'	Triangle in forewing crossed	**91**
18	R3 not undulate, or only slightly undulate (Fig. Li12)	**19**
18'	R3 undulate (Fig. Li11)	**21**
19	Two or more cubito-anal crossveins in hind wing	**26**
19'	One cubito-anal crossvein in hind wing	**27**
20	R3 undulate (Fig. Li11)	**21**
20'	R3 not undulate (Fig. Li12)	**22**
21	Discoidal field of forewing parallel sided or somewhat narrowed distally; sectors of arculus in forewing stalked (Fig. Li4); no supratriangular crossvein, usually only one cubito-anal crossvein in hind wing	***Orthemis*** (p. 117)
21'	Discoidal field of forewing widened distally; sectors of arculus in forewing not stalked (Fig. Li3), supratriangular crossvein usually present; usually two or more cubito-anal crossveins in hind wing	***Libellula*** (p. 110)
22	Arculus in hind wing between first and 2nd antenodals	**23**
22'	Arculus in hind wing opposite or distad of 2nd antenodal	***Cannaphila*** (p. 102)
23	Triangle in forewing free	**24**
23'	Triangle in forewing crossed	**25**
24	Costal side of triangle in forewing broken, usually near middle; CuP in hind wing usually separated from anal angle of triangle (Fig. Li6); 1-2 cell rows in discoidal field of forewing	***Nephepeltia*** (p. 116)
24'	Costal side of triangle in forewing straight; CuP in hind wing variable; 2-3 cell rows in discoidal field of forewing	**72**
25	Aspl in hind wing bent at level of heel (Figs. Li 7, 8)	**91**
25'	Aspl in hind wing straight or nearly so	**73**
26	Hind wing 20-26 mm	**27**
26'	Hind wing 29-35 mm	***Cannaphila*** (p. 102)
27	Males	**28**
27'	Females	**29**
28	Hamules 2-parted (Fig. Li20); 1-3 cell rows in discoidal field of forewing	***Erythrodiplax*** (p. 106)
28'	Hamules not 2-parted; 1 cell row in discoidal field of forewing	***Elga*** (p. 103)

29   Vulvar lamina scoop-shaped, one fourth as long as segment 9 or longer, and projecting ventrad (Fig. Li21); segment 10 not prolonged distally on ventral side
....***Erythrodiplax*** (p. 106)
29'   Vulvar lamina poorly developed, little more than a bilobed thickening of the caudal margin of the eighth sternite, and not projecting ventrad; segment 10 usually prolonged distally on ventral side to about opposite apex of superior appendages; nine antenodals in forewing..***Elga*** (p. 103)
30   Anal loop open posteriorly and extending to wing margin (Fig. Li7)
....***Tholymis*** (p. 124)
30'   Anal loop closed posteriorly and not extending to wing margin (Fig. Li8)........**31**
31   R3 undulate (Fig. Li11)..**32**
31'   R3 not undulate (Fig. Li12)..**39**
32   One bridge crossvein..**33**
32'   Two or more bridge crossveins..**38**
33   Mspl indistinct, or, if distinct, then usually 1 cell row between it and MA......**34**
33'   Mspl distinct, with 2 or more cell rows between it and MA, at least in forewing
....**36**
34   A transverse carina on segment 5 (Fig. Li15); R3 very undulate (Fig. Li11); base of hind wing noticeably broadened, with 8 or more rows of cells in anal field
....***Pantala*** (p. 119)
34'   No transverse carina on segment 5; R3 moderately to slightly undulate; base of hind wing not noticeably broadened with 2-5 rows of cells in anal field..........**35**
35   Discoidal field of forewing with two cell rows, at least for a distance of three or more cells; base of triangle in hind wing usually slightly proximad of arculus; triangle in forewing free or crossed; spines on outer angle of hind femur of male short, stout, and directed proximad....................................***Brechmorhoga*** (p. 101)
35'   Discoidal field of forewing with 3 cell rows, or rarely with 2 rows for a distance of one or two cells; base of triangle in hind wing usually opposite arculus; triangle in forewing crossed; spines on outer angle of hind femur of male variable..***Dythemis*** (p. 103)
36   Postnodal portion of wings hyaline; ends of pterostigma not parallel; triangle in hind wing free..***Paltothemis*** (p. 119)
36'   Postnodal portion of wings with some dark coloration; ends of pterostigma parallel; triangle in hind wing free or crossed..**37**
37   IR2a arising under proximal end of pterostigma; area between nodus and pterostigma largely black; triangle in hind wing crossed....***Pseudoleon*** (p. 122)
37'   IR2a arising under middle of pterostigma; wing color not as above
....***Libellula*** (p. 110)
38   Ends of pterostigma not parallel....................................***Paltothemis*** (p. 119)
38'   Ends of pterostigma parallel..***Libellula*** (p. 110)
39   CuP in hind wing arising at anal angle of triangle (Fig. Li5) or only slightly separated from it..**40**
39'   CuP in hind wing separated from anal angle of triangle (Figs. Li6)..........**74**
40   Triangle in forewing free..**41**
40'   Triangle in forewing crossed..**49**
41   One bridge crossvein..**42**
41'   Two or more bridge crossveins..**48**

42   Posterior lobe of prothorax not narrowed at base (Figs. Li13), and usually not erect or bearing a fringe of long hairs _____ **43**

42'   Posterior lobe of prothorax narrowed at base (Figs. Li14), and usually erect and bearing a fringe of long hairs _____ **72**

43   Mspl distinct in forewing _____ **44**

43'   Mspl not distinct in forewing _____ **45**

44   Subtriangle in forewing 1-celled; costal side of triangle in forewing more than half as long as proximal side; sectors of arculus in forewing not stalked   (Fig. Li100) _____ ***Idiataphe*** (p. 110)

44   Subtriangle in forewing 2- or 3-celled; costal side of triangle in forewing less than half as long as proximal side; sector of arculus in forewing stalked _____ **62**

45   IR2b well developed, with most of the cells above it elongated vertically (Fig. Li9); ends of pterostigma not parallel _____ ***Miathyria*** (p. 113)

45'   IR2b not developed, usually no vertically elongated cells in this section of wing; ends of pterostigma usually parallel _____ **46**

46   Nodus in forewing distad of middle of wing _____ **47**

46'   Nodus in forewing in approx. the middle of wing _____ ***Erythrodiplax*** (p. 106)

47   Discoidal field of forewing slightly widened distally ____ ***Brechmorhoga*** (p. 101)

47'   Discoidal field of forewing parallel-sided or narrowed distally
        _____ ***Macrothemis*** (p. 111)

48   Aspl in hind wing straight or nearly so; posterior lobe of prothorax narrowed at base, and usually erect and bearing a fringe of long hairs; costal side of triangle in forewing half as long as proximal side, or longer _____ **73**

48'   Aspl in hind wing bent at level of heel (Figs. Li7, 8); posterior lobe of prothorax not narrowed at base (Figs. Li13), and usually not erect or bearing a fringe of long hairs; costal side of triangle in forewing usually less than half as long as proximal side _____ ***Micrathyria*** (p. 114)

49   One bridge crossvein _____ **50**

49'   Two or more bridge crossveins _____ **68**

50   Arculus in forewing between first and second antenodals _____ **51**

50'   Arculus in forewing opposite or distad of second antenodal _____ **57**

51   Posterior lobe of prothorax narrowed at base (Figs. Li13), and usually erect and bearing a fringe of long hairs _____ **52**

51'   Posterior lobe of prothorax not narrowed at base (Figs. Li14), and usually not erect or bearing a fringe of long hairs _____ **60**

52   Aspl in hind wing bent at level of heel (Figs. Li7, 8) _____ **53**

52'   Aspl in hind wing straight or nearly so _____ **73**

53   Sectors of arculus in forewing stalked (Fig. Li4) _____ **54**

53'   Sectors of arculus in forewing not stalked (FIg. Li3) _____ ***Brachymesia*** (p. 100)

54   One cell row between IR3 and Rspl _____ **55**

54'   Two cell rows between IR3 and Rspl _____ **56**

55   Discoidal field of forewing parallel sided or narrowed distally; usually no lateral keel on segment 9 _____ ***Sympetrum*** (p. 122)

55'   Discoidal field of forewing slightly widened distally; a lateral keel on 9 _____ **82**

56   Discoidal field of forewing slightly widened distally; 12½ or more antenodals in forewing; costal side of triangle in forewing less than one-third as long as proximal side _____ ***Erythemis*** (p. 104)

56' Discoidal field of forewing parallel sided or slightly narrowed distally; 10½ or fewer antenodals in forewing; costal side of triangle in forewing more than one-third as long as proximal side ....*Sympetrum* (p. 122)

57   One cubito-anal crossvein in hind wing..**58**

57' Two or more cubito-anal crossveins in hind wing..**66**

58   Posterior lobe of prothorax narrowed at base (Figs. Li13), and usually erect and bearing a fringe of long hairs..**59**

58' Posterior lobe of prothorax not narrowed at base (Figs. Li14), and usually not erect or bearing a fringe of long hairs..**61**

59   Hind wing 20-28 mm; spines on hind and middle femora of both sexes slender, gradually increasing in length distally..*Erythrodiplax* (p. 106)

59' Hind wing 35-40 mm; spines on basal two-thirds of middle femur of male short, stout, blunt-tipped, and of approximately uniform length, those on distal third long and slender..*Rhodopygia* (p. 122)

60   Four cell rows in discoidal field of forewing; base of hind wing broad, with many rows of cells in anal field; ends of pterostigma not parallel
....*Tramea* (p. 124)

60' Two or three cell rows in discoidal field of forewing; pterostigma and base of hind wing variable..**61**

61   Two cell rows in discoidal field of forewing, at least for three cells..**62**

61' Three cell rows in discoidal field of forewing..**63**

62   Nodus in forewing distad of middle of wing; base of triangle in hind wing usually slightly proximad of arculus; spines on outer angle of hind femur of male short, stout, and directed proximad; vulvar lamina small, bilobed, and not projecting ventrad (Fig. Li22)..*Brechmorhoga* (p. 101)

62' Nodus in forewing in approximately the middle of wing; base of triangle in hind wing opposite arculus; spines on outer angle of hind femur of both sexes slender, gradually increasing in length distally, and not directed proximad; vulvar lamina at least one-fourth as long as segment 9 (often longer), rounded at apex, and projecting ventrad (Fig. Li21)..*Erythrodiplax* (p. 106)

63   Ends of pterostigma not parallel; anal field of hind wing several cells wide, the cells not arranged in rows parallel to A2..**64**

63' Ends of pterostigma parallel, or if not parallel, then 3-4 rows of cells in anal field of hind wing, the cells usually arranged in rows parallel to A2..**65**

64   One cell row between MA and Mspl; 1-2 cell rows between IR3 and Rspl
....*Tauriphila* (p. 123)

64' Two cell rows between MA and Mspl, at least in forewing; 2 cell rows between IR3 and Rspl..*Paltothemis* (p. 119)

65   Hamules 2-parted (Fig. Li20); vulvar lamina well developed, at least one-fourth as long as segment 9 and projecting ventrad (Fig. Li21), the apex in ventral view rounded; spines on outer angle of hind femur of both sexes slender, gradually increasing in length distally..*Erythrodiplax* (p. 106)

65' Hamules not 2-parted (Fig. Li19); vulvar lamina poorly developed, little more than a bilobed thickening of the caudal margin of the eighth sternite, not projecting ventrad; spines on outer angle of hind femur of male short and stout
....*Dythemis* (p. 103)

66   Base of triangle in hind wing distad of arculus; wing tips usually dark; vulvar
        lamina and sternites of segment 9 long, extending beyond apex of abdomen
        ....***Uracis*** (p. 124)

66'   Base of triangle in hind wing opposite arculus; wing tips hyaline; vulvar lamina
        half as long as segment 9, or less **67**

67   Hamules 2-parted (Fig. Li20); vulvar lamina scoop-shaped, one-fourth as long as
        segment 9 or longer, and projecting ventrad (Fig. Li21) ...***Erythrodiplax*** (p. 106)

67'   Hamules not 2-parted (Fig. Li19); vulvar lamina poorly developed, little more
        than a bilobed thickening of the caudal margin of the eighth sternite, and not
        projecting ventrad..***Dythemis*** (p. 103)

68   Posterior lobe of prothorax not narrowed at base (Figs. Li14), and usually not
        erect or bearing a fringe of long hairs..**69**

68'   Posterior lobe of prothorax narrowed at base (Figs. Li13), and usually erect and
        bearing a fringe of long hairs..**72**

69   Two cell rows between IR3 and Rspl; Mspl well developed; hind wing 43-48
        mm...***Paltothemis*** (p. 119)

69'   One cell row between IR3 and Rspl; Mspl indistinct; hind wing 18-35 mm.....**70**

70   Triangle in hind wing free, its base opposite arculus; 1-2 cubito-anal crossveins
        in hind wing; wing tips usually hyaline; vulvar lamina not extending beyond
        apex of abdomen..**71**

70'   Triangle in hind wing free or crossed, its base distad of arculus; 1-6 cubito-anal
        crossveins in hind wing; wing tips usually dark; vulvar lamina and sternite of
        segment 9 long, extending beyond apex of abdomen.................***Uracis*** (p. 124)

71   Two or more bridge crossveins in each wing.....................***Micrathyria*** (p. 114)

71'   At least one wing with only one bridge crossvein.............***Erythrodiplax*** (p. 106)

72   Aspl in hind wing straight or nearly so..**73**

72'   Aspl in hind wing bent at level of heel (figs. Li7, 8)...**91**

73   Discoidal field of forewing not widened distally, usually narrowed distally;
        wings often yellowish, or with brownish or black markings distal to triangle
        ....***Perithemis*** (p. 119)

73'   Discoidal field of forewing usually widened distally; wings hyaline, often with a
        small brownish or blackish basal spot in hind wing which does not extend
        beyond triangle..***Planiplax*** (p. 120)

74   One bridge crossvein..**75**

74'   Two or more bridge crossveins..**89**

75   Arculus in hind wing between first and second antenodals...........................**76**

75'   Arculus in hind wing opposite or distad of second antenodal........................**84**

76   Tooth on tarsal claw small and near tip of claw, or very small, little more than a
        notch (Fig. Li18); hamules not 2-parted (Fig. Li19)...**77**

76'   Tooth on tarsal claw usually normally developed (Figs. Li17); hamules variable
        ....**78**

77   Sectors of arculus in forewing stalked (Fig. Li4).................***Oligoclada*** (p. 117)

77'   Sectors of arculus in forewing not stalked (Fig. Li3)..............***Planiplax*** (p. 120)

78   Posterior lobe of prothorax not narrowed at base (Figs. Li14), and usually not
        erect or bearing a fringe of long hairs..**79**

78'   Posterior lobe of prothorax narrowed at base (Figs. Li13), and usually erect and
        bearing a fringe of long hairs..**80**

79   Nodus in forewing distad of middle of wing.....................M***acrothemis*** (p. 111)

79'   Nodus in forewing in approx. the middle of wing ....*Erythrodiplax* (p. 106)
80   IR2a arising below proximal third of pterostigma, or farther proximad............**81**
80'   IR2a arising below middle third of pterostigma.................*Brachymesia* (p. 100)
81   One cell row between IR3 and Rspl...**82**
81'   Two cell rows between IR3 and Rspl...**83**
82   Spines on basal half or two-thirds of outer angle of hind femur short and about
        equal in length, with 3-4 large spines on distal half or third; hind wing 26 mm or
        more, usually over 30 mm; 3 cell rows in discoidal field of forewing
        ....*Erythemis* (p. 104)
82'   Spines on outer angle of hind femur gradually increasing in length distally, or all
        short except the last, hind wing 29 mm or less, usually less than 26 mm; 2-3
        rows in discoidal field of forewing.........................*Erythrodiplax* (p. 106)
83   Hind wing 22-29 mm.......................................*Erythrodiplax* (p. 106)
83'   Hind wing 35-41 mm...*Erythemis* (p. 104)
84   Tooth on tarsal claw usually as long as tip of claw, or longer (Figs. Li16), except
        in *tesselata* & *aurimaculata*; spines on outer angle of hind femur of male short
        and stout and usually directed proximad; nodus in forewing distad of middle of
        wing...*Macrothemis* (p. 111)
84'   Tooth on tarsal claw shorter than tip of claw (Figs. Li17); spines on outer angle
        of hind femur and nodus in forewing variable.....................................**85**
85   Nodus in forewing distad of middle of wing...**86**
85'   Nodus in forewing at approximately the middle of wing **87**
86   One cell row in discoidal field of forewing; costal side of triangle in forewing
        broken...*Elga* (p. 103)
86'   Two or three cell rows in discoidal field of forewing; costal side of triangle in
        forewing straight...*Dythemis* (p. 103)
87   One cell row in discoidal field of forewing; tooth on tarsal claw very small, little
        more than a notch (Fig. Li18)..*Elga* (p. 103)
87'   Two cell rows in discoidal field of forewing; tooth on tarsal claw normally
        developed (Figs. Li17)...**88**
88   One row of cells in anal field of hind wing; triangle in forewing free; inferior
        angle of male superior appendages located at one-third to three-fifths the length
        of appendage, the tips of superior appendages long and often upturned at apex
        ....*Anatya* (p. 100)
88'   Two or more rows of cells in anal field of hind wing; triangle in forewing free or
        crossed; inferior angle of male superior appendages located at about three-
        fourths the length of appendage, the tips of superior appendages not long and
        upturned at apex...*Erythrodiplax* (p. 106)
89   Two or more bridge crossveins in each wing...**90**
89'   One or two wings with only one bridge crossvein.....................................**91**
90   Sectors of arculus in forewing stalked (Fig. Li4)...........*Micrathyria* (p. 114)
90'   Sectors of arculus in forewing not stalked (Fig. Li3)..........*Planiplax* (p. 120)
91   IR2a arising below proximal end of pterostigma, or farther proximad; sectors of
        arculus in forewing stalked (Fig. Li4); hamules 2-parted (Fig. Li20)
        ....*Erythrodiplax* (p. 106)
91'   IR2a variable; sectors of arculus in forewing not stalked (Fig. 3); hamules not 2-
        parted (Fig. Li19)...**92**

92   IR2a arising below proximal end of pterostigma, or farther proximad; wings
        hyaline, with a small basal spot in hind wing which does not extend beyond
        triangle .... ***Planiplax*** (p. 120)
92'   IR2a arising below middle of pterostigma (Fig. Li24); wing color variable, often
        with yellowish or dark areas distal to triangle .... ***Brachymesia*** (p. 100)

After: BORROR (1945)

Anatya

Anatya guttata

Males of this small sized dragonfly are characterized by their distinctive anal
appendages, which are long, white, and forming a sine curve when viewed laterally.
Thorax and abdomen predominantly blackish with pale greenish-blue markings. The
species generally inhabits ponds, although adults may be encountered on any small
standing water as well as in semi-open areas inside forests (MAY, 1979; MICHALSKI,
1988).

Brachymesia

These are medium sized dragonflies having their abdomen swollen at segments 2 and 3.
They occur at ponds and lakes, also brackish ones, where males patrol territories
(DUNKLE, 1989).

1   Black on abdominal dorsum confined to mid-dorsal spot or stripe on segments 8-
        9, or still more reduced; ground-colour of abdomen red (male) or luteous
        (female), face and labrum reddish or luteous; forewings with 8-10 antenodals
        (wings as in Fig. Li24, Pl. XXI); males with the anterior lamina not more
        prominent than the hamule, its apical fifth bilobed .... ***furcata***
1'   Black on abdominal dorsum forming a median band from the apex of segment 3-
        9; ground colour of abdomen in younger individuals reddish- or yellowish-
        brown; face and labrum luteous, the latter sometimes edged with black;
        forewings with 10-12 antenodals; males with the anterior lamina more prominent
        than hamule or genital lobe, its apical half bilobed .... ***herbida***

After: CALVERT (1908)

Brechmorhoga

Plate XIX, p. 105

Medium sized dragonflies (abdomen 35-40 mm). Eyes blue or bluish green. Thorax generally dark with pale greenish yellow markings (sometimes predominantly pale). Abdomen in males widened at segment 7 and 8, predominantly black with pale markings, bearing a conspicuous pale spot on either side of dorsum of segment 7. They live on streams, preferably in forested areas, where males can be seen patrolling along the stream edge (ALCOCK, 1989). However, NOVELO GUTIÉRREZ (1989) found *B. praecox* to inhabit mainly more open habitats, e.g. broad shallow streams and rivers with little vegetation shading the water. There the larvae could be found in areas with higher current between gravel and small rocks. Adults may be found also at some distance to water on clearings in the forest, along trails, and even on slightly wooded pastures.

1	Males with two posttriangular rows of cells in hind wing, sometimes with one cell reaching from MA to CuP, followed immediately by two rows; genital lobe as prominent as anterior lamina; superior appendages with inferior denticles, but no tooth; Females with the 11th abdominal tergite half as long as abdominal appendages	**2**
1'	Males with one posttriangular row of cells in hind wing, at least for three cells; genital lobe half or less than half as prominent as anterior lamina (Fig. Li26); superior appendages with an inferior tooth; Females with the 11th abdominal tergite as long as the abdominal appendages	***nubecula***
2	Hind wings with 2 rows of cells between A3 and the wing margin posterior to the membranule	**3**
2'	Hind wings with 3 rows of cells between A3 and the wing margin posterior to the membranule; female with vulvar lamina bilobed at its extremity, interval separating the lobes wider than deep	**4**
2''	Hind wings with 3 rows of cells between A3 and the wing margin posterior to the membranule followed by 2 rows (male) or 3 rows (female); female with vulvar lamina bilobed, interval separating the lobes narrowly U-shaped, deeper than wide	***tepeaca***
3	Abdominal segment 3 with the longitudinal green stripe on each side of the dorsum not confluent with the transverse basal green line; labrum usually margined with brown or black, especially at the sides; males with hamule strongly curved throughout its length; the metallic blue of the superior and anterior surfaces of the frons extending also on to the lateral surfaces thereof; females with the lobes of the vulvar lamina shorter than the width of the interval separating them, this interval much wider at the apices of the lobes than deep; abdomen subequal in length to the hind wing	***vivax***
3'	Abdominal segment 3 with the longitudinal green stripe on each side of the dorsum confluent with the transverse basal green line; labrum pale, unmarked; males with hamule not so strongly curved, nearly straight in the middle; the metallic blue of the superior and anterior surfaces of the frons not extending on to the lateral surfaces thereof; females with the lobes of the vulvar lamina longer	

than the width of the interval separating them, this interval narrower at the apices of the lobes than deep; abdomen longer than the hind wing .... ***praecox***

4   Hamule of the male not equally curved throughout, but almost straight in its middle, bent near the apex, which is less acute and less tapering than in *rapax* (Fig. Li27); vulvar lamina similar to that of *vivax*...................................***pertinax***

4'   Hamule of the male strongly and equally curved throughout and tapering to an acute apex (Fig. Li25); vulvar lamina rounded at apex, slightly divided, but the two divisions in contact with each other throughout...................................***rapax***

After: CALVERT (1908)

Cannaphila

Plate XIX, p. 105

Comparatively small sized dragonflies (abdomen 25-30 mm). Young individuals with pale areas on abdomen reddish-yellow (*insularis* and *vibex*) or brownish (*mortoni*), becoming either dark grey with pruinose blue areas (*vibex* and *insularis*), bright red (*mortoni*) or remaining reddish-yellow (some *vibex*) (CALVERT, 1908; DONNELLY, 1992). Females with a foliation on 8th abdominal segment. *C. vibex* generally occurs in open seepage areas (DUNKLE, in litt.), whereas *C. insularis* and *C. mortoni* are found in wooded swamps or along streams at lower elevations (DONNELLY, 1992).

1   Midrib of anal loop arises proximal to the second cubital crossvein (Fig. Li29); transverse carina of third abdominal segment of female curves into lateral carina ....***vibex***

1'   Midrib of anal loop arises distal to second cubital crossvein; transverse carina of third abdominal segment meets lateral carina at right angle...................**2**

2   Posterior hamule on second abdominal segment of male with prominent hook; anal loop with 6-9 cells (Fig. Li31); mature male with red abdomen, old males with reddish black abdomen, foliation on 8th abdominal segment of female 0.54-0.62 mm wide; female abdomen with no lateral yellow line...................***mortoni***

2'   Posterior hamule on second abdominal segment of male with small hook; anal loop with 9-15 cells (Fig. Li30); mature males with pruinose black abdomen; foliation on 8th abdominal segment of female 0.75-0.91 mm wide; female abdomen with lateral yellow stripe on basal and middle abdominal segments ....***insularis***

After: DONNELLY (1992)

Dythemis

Plate XIX, p. 105

Small to medium sized libellulids allied to *Macrothemis* and *Brechmorhoga*. However, males are distinguished by having the abdomen not as widened at segment 7 and 8 as in the latter genera. They are generally found on streams.

Males

1   Anterior lamina of the male less prominent than genital lobe (Fig. Li33) **2**

1'   Anterior lamina of the male more prominent than genital lobe (Fig. Li32); thorax dark reddish brown, no pale antehumeral stripes; wings uncoloured, or with a very faint yellowish tinge at extreme base out to less than half-way to the submedian crossvein; hind wing with 3-4 rows of cells between A3 and the anal angle ....*cannacrioides*

                                                    now = *Elasmothemis cannacrioides*

2   Thorax brown to black, each side with a pale antehumeral stripe one-third to one-fifth as wide as the dark color which separates it from the mid-dorsal carina **3**

2'   Thorax reddish, no pale antehumeral stripes; wings reddish brown (male) or yellowish (female) at base out to the level of the middle of the internal triangle on forewing, to distal angle of triangle of hind wing; hind wing with 6-7 rows of cells between A3 and the anal angle...*maya*

3   Frons yellow to yellowish-red; thorax dark brown; yellow basal spot of wings reaching cubital crossvein (male) or further distad (female); wings light yellowish with more or less distinct dark apices; abdominal segments 4-7 with yellowish lateral stripes more than half as long as segment......................*sterilis*

3'   Frons blue-green to blue metallic; thorax black with green-metallic shine; basal spot of wings mostly not reaching the cubital crossvein; wings hyaline or light greyish to yellowish tinged, apices highly variable; abdominal segments 4-7 with very small greenish spots on lateral side, less than half as long as segment ....*multipunctata*

Derived from: CALVERT (1908), RIS (1918), WESTFALL (1988)

Elga

Elga leptostyla

A small sized libellulid dragonfly (abdomen about 20 mm in length). Thorax olive green, sometimes almost brown, with two yellowish to greenish markings on mesepisternum close to the humeral suture. Abdomen black with yellow markings, except segments 1-3 and 10, which are almost completely yellowish (MACHADO, 1954). DE MARMELS (1992) stated this species to be „typical for small creeks in the forest where males use to sit on sticks in the sun."

Erythemis

Species of this genus of medium sized dragonflies are remarkably variable in general appearance and coloration. Thorax moderately robust, abdomen stout to slender, inflated or not at the base, never distinctly dilated apically. A striking difference within the genus is the color of adults, especially of males. Some have more or less of the thorax, but especially the abdomen, brilliant red, while another group includes species which are largely greenish or yellowish as tenerals and become dark bluish pruinose with age, others becoming deep black with age. Lateral margins of females not expanded (WILLIAMSON, 1923b).

1   Two cell rows between IR3 and Rspl; large, green and black species, abdomen slender .... ***vesiculosa***

1'   One cell row between IR3 and Rspl **2**

2   Lateral and ventral carinas on abdominal segment 3 at the apex, measured along the apical carina, separated by 0.8 mm or less; the distance between them, opposite the meeting point of the lateral and median transverse carinas, more than one and one-half times the distance between them at apex **3**

2'   Lateral and ventral carinas on abdominal segment 3 at the apex, measured along the apical carina, separated by 1 mm or more; the distance between them, opposite the meeting point of the lateral and median transverse carinas, equal to one and one-half times or less the distance between them at apex **6**

3   Abdomen less than 30 mm long, shorter than the hind wings; dorsum of thorax distinctly patterned, paler above, bordered on either side with black .... ***credula***

3'   Abdomen 30 mm or longer **4**

4   Tibiae, and to a lesser extent the femora, largely dark brown or black; abdomen and hind wings about equal in length; female lamina trough-shaped, projecting ventrad, the apex not distinctly bent and not directed caudad **5**

4'   Legs largely pale; hind wings longer than the abdomen; thorax not distinctly patterned paler and darker, yellowish or greenish to red insects; female lamina distinctly bent so the apical lobe is directed caudad, and in lateral view there is a deep posterior emargination between basal lobes and apex .... ***carmelita***

5   Dorsum of abdominal segments 5-10 predominantly dark or black; dorsum of thorax pale above, bordered on either side with a black antehumeral band or the thorax black; female lamina from base of basal lobe to apex 1.5 mm ...... ***plebeja***

5'   Dorsum of abdominal segments 5-10 predominantly pale; dorsum of thorax not distinctly patterned paler and darker; female lamina from base of basal lobe to apex 0.75 mm .... ***haematogastra***

6   Labrum, face and frons pale colored, green and yellow; wing bases unmarked or, at the most, hind wings basally tinged with yellow; male with ventral teeth on the superior appendages extending posteriorly beyond the level of the apex of the inferior appendages .... ***simplicicollis***

6'   Labrum, face and frons not as above; base of hind wings distinctly dark or at least (some females) tinged with yellow; male with ventral teeth on superior appendages extending posteriorly to about the same level as the apex of the inferior apendage or not that far **7**

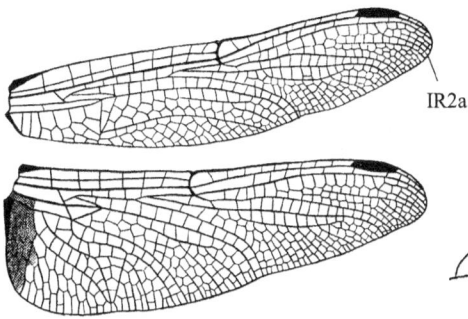

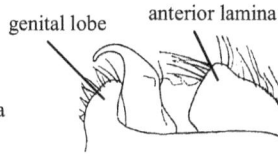

genital lobe   anterior lamina

Plate XIX

Li25

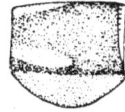

C. insularis

IR2a

Li26

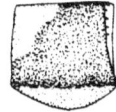

C. mortoni

Fig. Li24: Wings of *Brachymesia furcata*

Fig. Li25-27: *Brechmorhoga*, male accessory genitalia, left lateral view; **Li25** - *B. rapax*, **Li26** - *B. nubecula*, **Li27** - *B. pertinax*

Li27

Fig. Li28: *Cannaphila*, 8th abdominal segment of female in lateral view

Li29   Li30   Li31

Fig. Li29-Li31: *Cannaphila*, base of hind wing, **Li29**: *C. vibex*; **Li30**: *C. insularis*; **Li31**: *C. mortoni*

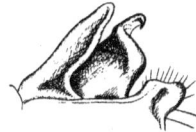

Li32

Li33

Fig. Li32 & 33: Lateral view of male accessory genitalia; **Li32**: *Dythemis* (*Elasmothemis*) *cannacrioides*; **Li33**: *D. multipunctata*

Li34 *abjecta*

Li35 *andagoya*

Li36 *andagoya*

Li37 *berenice*

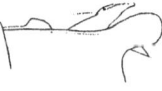

Li38 *berenice*

Li39 *berenice*

Li40 *connata*

Li41 *connata*

Li42 *connata*

Li43 *famula*

Li44 *famula lativittata*

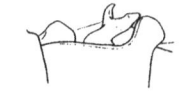

Li45 *f. lativittata*

Li46 *fervida*

Li47 *fervida*

Li48 *fusca*

Li49 *fusca*

Li50 *kimminsi*

Li51 *unimaculata*

Li52 *unimaculata*

Li53 *unimaculata*

Figs. Li34-Li53: *Erythrodiplax*, lateral view of male accessory genitalia (of penis in Li36 & 39)

7   Dark basal area on hind wing reduced, not reaching the cubito-anal crossvein; thorax distinctly patterned with a median rectangular pale (usually yellow) area above, bordered with black, to entirely black in older males which have abdominal segments 4-10 bright red; female vulvar lamina in antero-ventral view semicircular..***peruviana***

7'   Dark basal area on hind wing reaching beyond the level of the cubito-anal crossvein; thorax not as above; female lamina in antero-ventral view not semicircular..**8**

8   Thorax largely dark or black; abdomen black in old males, in others largely dark with conspicious large basal or subbasal pale (yellow) areas, only narrowly separated by black in the median line of segments 4 and 7; male with the apex of the inferior appendage reaching beyond the level of the ventral teeth on the superior appendage; female lamina in antero-ventral view nearly as long as wide, rounded triangular..***attala***

8'   Thorax pale yellowish or greenish to dark red, without dark markings; abdomen largely pale, yellowish to bright red; male with the ventral teeth on the superior appendages extending posteriorly to about the same level as the apex of the inferior appendage; female lamina in antero-ventral view trilobed......***mithroides***

Derived from: MICHALSKI (1988), WILLIAMSON (1923b)

Erythrodiplax
Plates XIX (p. 105) & XX (p. 109)

Small to medium sized dragonflies (abdomen 12-33 mm). Wing coloration highly variable in different species and in different individuals of the same species. In most species wings with some colored markings, varying in color from faint yellowish to dark reddish brown or black. Wing-spotted forms usually with a basal spot, at least in hind wing, sometimes also with crossbands. Abdomen usually slender and cylindrical, segment 1-3 of male swollen, 4 and 5 more slender, the remaining segments sligthly more robust. Color of abdomen highly variable in different species and in the same species with age and source locality. Teneral species usually with abdomen yellowish or yellowish brown, in mature males yellowish brown to bluish black, often covered with a bluish pruinescence; sometimes also red-coloured.

They are common at standing waters like pools and swamps, but also at small rivers in open areas and often perch low over the ground.

1   Two cell rows between IR3 and Rspl; CuP in hind wing arising at anal triangle; adult male usually with a conspicious dark band across wings.............................**2**

1'   Usually one cell row between IR3 and Rspl; if two rows, then adult male is without a conspicious dark band across wings..**3**

2   Usually one cell row between MA and Mspl; adult male and homochromatic female with a brown or black band across each wing between nodus and pterostigma...**umbrata**

2'   Usually two cell rows between MA and Mspl; adult male and homochromatic
        female with wings brown or black from base or from triangle to halfway between
        nodus and pterostigma..***funerea***
3   Lateral keel on segment 2 bent forward sharply at a right angle; posterior lobe of
        prothorax narrowed basally and distinctly widened distally; penultimate spine on
        outer angle of hind femur three-fourths as long as ultimate spine or longer,
        hamules very robust, as wide in profile as genital lobe...........................***castanea***
3'   Lateral keel on segment 2 bent forward in a rounded curve; posterior lobe of
        prothorax variable, usually not widened distally, penultimate spine on outer
        angle of hind femur usually half as long as ultimate spine, or less; hamules
        slender, in profile narrower than genital lobe...**4**
4   Males..**5**
4'   Females..**14**
5   Genital lobe usually upright and not overlying posteriorly, truncate, with distal
        edge nearly or quite straight (e.g. Figs. Li46, 50, 51); wing color variable; penis
        with paired internal lobes..**6**
5'   Genital lobe rounded, or, if distal edge is straight, then lobe is strongly overlying
        posteriorly (e.g. Figs. Li40, 45, 49); wings usually with a small basal spot only,
        rarely with tips brown or black; penis variable, sometimes without paired
        internal lobes...**8**
6   Frons of adult metallic blue, at least in part...**7**
6'   Frons of adult reddish or brownish, with no blue..................................***fervida***
7   Basal spot in hind wing extending to distal angle of triangle or slightly farther,
        over nearly the entire width of wing base; basal spot in forewing extending to
        first, second, or third antenodal..***unimaculata***
7'   Basal spot in hind wing extending to first or second antenodal, base of A2 and
        anal angle of wing, frequently bordered by an opal band; spot usually narrowed
        posteriorly..***kimminsi***
8   CuP in hind wing arising from anal angle of triangle or only slightly separated
        from it...**9**
8'   CuP in hind wing distinctly separated from anal angle of triangle...............**12**
9   Genital lobe small, more or less rounded distally, and not conspiciously
        overlying posteriorly (Figs. Li43-45); posterior lobes of prothorax distinctly
        narrowed and rounded distally; hind wings with basal spot variable in size and
        color, but with spot in forewing usually extending nearly as far distad as that in
        hind wing...**10**
9'   Genital lobe variable, usually distinctly overlying posteriorly (e.g. Li38, 49);
        frons in adult bluish black or reddish brown, wing tips usually hyaline; basal spot
        in hind wing extending at most to base of triangle, usually less...................**12**
10   Basal spot in hind wing extending to the second, third, fourth, or fifth antenodal,
        usually completely covering triangle; frons, thorax, and abdomen in adult
        specimens reddish brown..***famula***
10'   Basal spot in hind wing extending to third antenodal and anal angle of triangle,
        or less; frons varying in color from a yellowish brown in teneral specimens to
        bluish black in adults; thorax and abdomen in adult specimens bluish black, the
        thorax with a greenish lateral stripe...***famula lativittata***
11   Terminal segment of penis 1.29 to 1.66 mm long.................................***abjecta***
11'   Terminal segment of penis 0.90 to 1.28 mm long.....................................**12**

12   Frons bluish black..*connata*
12'   Frons red or reddish brown, basal spot in hind wing dark red or reddish brown, and variable in size but usually extending at least to base of A2, with distal margin more or less rounded...*fusca*
13   Penis with paired internal lobes and lacking a posterior lobe (Fig. Li39); frequently two cell rows between IR3 and R3, at least for a short distance ....*berenice*
13'   Penis without paired internal lobes, usually with a well-developed posterior lobe (Fig. Li36); only one cell row between IR3 and R3..........................*andagoya*
14   Vulvar lamina not more than half as long as segment 9.............................**15**
14'   Vulvar lamina more than half as long as segment 9................................**20**
15   Vulvar lamina one-third as long as segment 9, or less, in profile narrowed at apex (Fig. Li57); posterior lobe of prothorax rounded and narrowed distally ....*famula*
15'   Vulvar lamina usually half as long as segment 9; if vulvar lamina is less than half as long as segment 9, then apex is not narrowed but broadly rounded in profile, and posterior lobe of prothorax is somewhat quadrate and not narrowed distally..**16**
16   Vulvar lamina in profile somewhat triangular, with sides nearly or quite straight and apex pointed (Fig. Li63)..**17**
16'   Vulvar lamina in profile with sides, at least dorsoposterior side, convexly curved, and apex pointed or somewhat rounded (Fig. Li56)............................ *connata*
17   Frons partly or entirely metallic blue..**18**
17'   Frons greenish, brownish, or yellowish, with no blue...............................**19**
18   Dorsoposterior edge of vulvar lamina in profile usually at almost a right angle to lateral margin of segment 9; tip of vulvar lamina but little or not at all bent caudad (Fig. Li63); blue frons usually occupying half the area of front of frons, or more..*unimaculata*
18'   Dorsoposterior edge of vulvar lamina in profile usually at a 60° angle or less to lateral margin of segment 9; tip of vulvar lamina bent slightly caudad (Fig. Li61); blue of frons usually restricted to that portion immediately below median ocellus, and occupying half the area of front of frons, or less...............*kimminsi*
19   Lateral spot on segment 7 usually extending dorsad to dorsal carina; dorsoposterior edge of vulvar lamina in profile usually at almost a right angle to lateral margin of segment 9; tip of vulvar lamina but little or not at all bent caudad (Fig. Li58)...*fervida*
19'   Lateral spot on segment 7 usually not extending dorsad to dorsal carina; dorsoposterior edge of vulvar lamina in profile usually at a 60° angle or less to lateral margin of segment 9; tip of vulvar lamina bent slightly caudad (Fig. Li61) ....*kimminsi*
20   Vulvar lamina in profile somewhat triangular, with dorso-posterior margin nearly or quite straight..*berenice*
20'   Vulvar lamina in profile not definitely triangular, with dorso-posterior margin convexly curved...**21**

Plate XX

Li54 *abjecta*

Li55 *andagoya*

Li56 *connata*

Li57 *famula lativittata*

Li58 *fervida*

Li59 *funerea*

Li60 *fusca*

Li61 *kimminsi*

Li62 *umbrata*

Li63 *unimaculata*

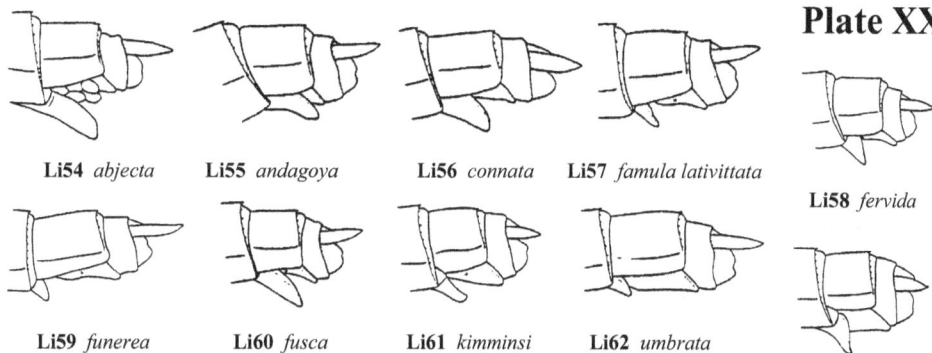

Figs. Li54-Li63: *Erythrodiplax*, lateral view of apex of female abdomen

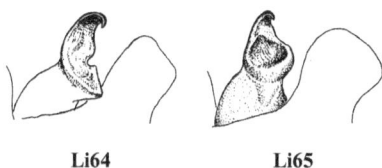

Li64

Li65

Figs. Li64 & 65: *Libellula*, lateral view (above) and ventro-lateral view (below) of male hamule
Li64: *Libellula foliata*; **Li65**: *L. mariae*

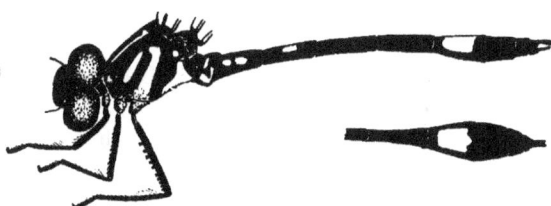

Fig. Li66: *Macrothemis aurimaculata*, male

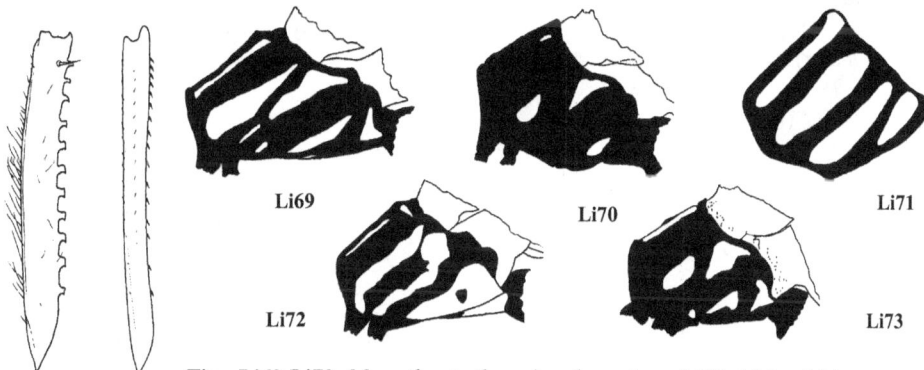

Li69

Li70

Li71

Li72

Li73

Figs. Li69-Li73: *Macrothemis*, thoracic color pattern; **Li69**: *M. hemichlora*; **Li70**: *M. imitans*, **Li71**: *M. inequiunguis*; **Li72**: *M.. musiva*, **Li73**: *M. pseudimitans*

Li67   Li68

Figs. Li67 & 68: *Macrothemis*, hind femur of male; **Li67** - *M. pseudimitans*, **Li68** - *M. ludia*

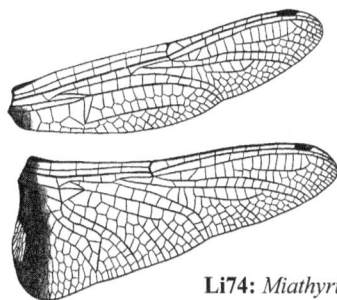

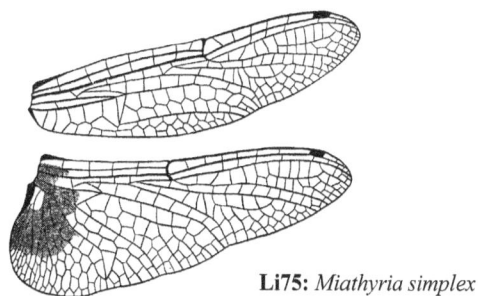

Li74: *Miathyria marcella*

Li75: *Miathyria simplex*

21   CuP in hind wing usually separated from anal edge of triangle; last antenodal in forewing complete or incomplete; no dark antehumeral stripe on thorax
....*andagoya*
21'   CuP in hind wing usually arising at anal angle of triangle or only slightly separated from it; last antenodal in forewing usually incomplete; usually a dark antehumeral stipe on thorax..**22**
22   Vulvar lamina as long as segment 9, or longer (Fig. Li54).....................*abjecta*
22'   Vulvar lamina shorter than segment 9 (Fig. Li56)..............................*connata*

After: BORROR (1942)

Idiataphe

These are small and slender, metallic black or brown dragonflies occuring at reedy ponds and lakes, where the males perch on the tips of emergent grasses, usually well out from shore. From there they make erratic patrol flights low over water (DUNKLE, 1989; MICHALSKI, 1988).

1   Forewing antenodal crossveins in male 6 ½, postnodals 5, usually 7 ½ and 6 in females; male has 2 anterior tufts of bristles on anterior lamina which are not on conical bases...*amazonica*
1'   Forewing antenodals in male 7 ½, postnodals 6, usually 8 ½ and 7 in females; male bristle tufts set on conical bases...*cubensis*

After: DUNKLE (in litt.)

Libellula
Plate XX, P. 109

Medium sized, stout bodied dragonflies. The females have lateral flaps on the margins of segment 8 which can be used as a scoop to throw drops of water containing eggs onto the bank.

1   Pterostigma 3.5 - 4.5 mm long on forewing..**2**
1'   Pterostigma 5 - 7 mm long on forewing; abdomen brilliant red, less intense in females..**3**
2   Dark basal wing markings present, reaching as far distad as the triangle; hamule forming a broad, cuplike structure with a broad, medially raised portion, posterior margin of the „cup" roughly convex (Fig. Li65)...................*mariae*
2'   Wings without such conspicious dark markings, somewhat smoky and sometimes with a slight flavescent tinge along the second series of antenodal and postnodal crossveins; posterior margin of the hamule with a distinct U-shaped cleft (Fig. Li64)..*foliata*

3   Wings without a distinctly colored basal area, usually somewhat smoky, especially at the apex (but sometimes smoky yellow throughout); pterostigma dark brown..*herculea*

3'   Wings with distinct coloration extending to the nodus; pterostigma yellowish to reddish brown..**4**

4   Wings yellowish brown at base for entire width, or only as far posteriorly as the submedian space; pterostigma yellow to red; frons without tubercles; outer branch of hamule with the distance from anterior to posterior margin of the ventral face less than half as great as from the mesal to the lateral, the margins not raised to form a ridge, inner branch long.......................................*croceipennis*

4'   Wings with a large orange basal area; pterostigma dark reddish brown; frons with tubercles present; outer branch of hamule with the distance from anterior to posterior margin of the ventral face almost as great as the distance from mesal to lateral, and the margins elevated to form a distinct ridge, inner branch short, terminating in a black recurved hook, directed laterad..........................*gaigei*

Compiled from: GARRISON (1973, 1992), GLOYD (1938)

Macrodiplax

Macrodiplax balteata

This is a medium sized dragonfly with a large rounded black spot at the base of each hindwing. Mature males have an entirely black face and abdomen, whereas females or juvenile males have a white face, a gray thorax with an irregular brown W on each side, abdominal segments 1 to 7 dull yellow, and segments 8 to 10 black. The species is typically found near the coast in brackish or mineralized waters. Males usually perch far from shore on emergent stems, from where they make patrols low over water with some hovering (DUNKLE, 1989).

Macrothemis

Plate XX, p. 109

Medium sized dragonflies similar to *Brechmorhoga*. Thorax dark with distinct pale greenish yellow markings, abdomen predominantly black. Males have abdomen widened at segments 7 and 8, with a conspicuous pale spot on dorsum of segment 7. They typically feed in high sustained flight. Males patrol edges of the current of rivers and streams. They sometimes perch flat on a rock or leaf like a gomphid (DUNKLE, in litt.).

Males

1   Tooth of the tarsal claw much shorter than the tip of the claw itself (Fig. Li17)
....**2**

1' Tooth of the tarsal claw as long as or longer than the tip of the claw itself (Fig.
Li16)..**3**

2   Pale orange, undivided spot on abdominal segment 7 occupies more than ¾ of
tergum (Fig. Li66); frons iridescent purple, with small yellow spots on ventral-
lateral corners; superior appendages in lateral view gradually thickened, with
maximum at 70% of length, the distal portion tapering, without a distinct,
prominent ventral tooth..***aurimaculata***

2' Paired rounded yellow-green spots on segment 7 occupying less than half of
tergum; frons centrally iridescent purple but broadly bordered laterally and
ventrally with yellow; superior appendages in lateral view with parallel sided
proximal portion, abruptly thickened at 40-50 % of length, with large, many
pointed teeth, and with distal portion parallel sided; thorax as in Fig. Li71
....***inequiunguis***

3   Superior appendages with apices bent distinctly downward and outward; femora
of hind legs distinctly curved..***nobilis***

3' Superior appendages with apices not bent distinctly downward and outward;
femora of hind legs usually nearly straight...**4**

4   Abdomen not much widened at segments 7 and 8.....................................**5**

4' Abdomen greatly widened at segments 7 and 8...**7**

5   Abdomen mostly dark with pale markings; sides of thorax with three pale green
stripes on dark ground (Fig. Li72)...**6**

5' Abdomen mostly cream to olive brown with dark markings on segments 1-3,
segments 4-8 largely pale or at least with extensive pale stripes bordering lateral
and median carinae..***inacuta***

6   Flexor surface of hind femur with row of short, stout teeth which are
subtriangular or curved proximally almost from their base, with distal margin of
each tooth oblique to long axis of femur (Fig. Li68); abdomen not markedly
widened on distal segments...***musiva***

6' Flexor surface of hind femur with short, stout teeth which are subquadrate, with
distal margin of each parallel to long axis of femur and proximal corner bent
proximally (Fig. Li67)...***extensa***

7   Superior appendages in lateral view with a distinct, ventral tooth at little more
than half their length, without additional denticles, apices in dorsal view nearly
straight and parallel, at most very slightly divergent; thorax as in Fig. Li70
....***imitans***

7' Superior appendages in lateral view without very prominent ventral tooth near
1/2 their length, in dorsal view usually not as above...................................**8**

8   Superior appendages with subterminal tooth or apical expansion, in lateral view
widest near apex, without ventral denticles..***delia***

8' Superior appendages without subterminal tooth or apical expansion, in lateral
view widest at 1/2 to 3/4 length, with series of 4 or more small ventral denticles
at or preceding widest point..**9**

9   Hind femur with subquadrate teeth of flexor surface large and robust; tibiae tan
to medium brown, distinctly paler than external surface of femora; lateral pale

markings on thorax consisting of 4-5 widely separated spots (Fig. Li73)
....*pseudimitans*

9'   Hind femur with subquadrate teeth of flexor surface relatively small; tibiae dark
      brown or black, not distinct from external surface of femora............................**10**

10   Mesepimeral pale area extending about 2/3 length of sclerite, metepisternal pale
      stripe constricted or very narrowly devided just behind spiracle; superior
      appendages in lateral view with distinct ventral angulation preceded by 6-10
      small denticles...*fallax*

10'   Mesepimeral stripe extending nearly full length of sclerite, metepisternal pale
      stripe broadly continuous behind spiracle (Fig. Li69); superior appendages
      rounded ventrally, without distinct ventral angulation, curved markedly upward
      and outward just beyond 1/2 length, bend preceded by several relatively coarse,
      sharp denticles...*hemichlora*

Compiled from: CALVERT (1898, 1908), MAY (1998)

Miathyria

Plate XX, p. 109

These medium sized dragonflies are generally associated with floating water plants.
They are similar to *Tramea* but are smaller and have a different arrangement of wing
veins. (DUNKLE, 1989).

1   Rspl forming a loop with IR3, enclosing 5-7 cells in the forewings, 6-7 cells in
      the hind wings; hind wings with a basal brown area, reaching posterior margin of
      wing, stopping at Aspl or, in a few females, reaching to the triangle (Fig. Li74);
      frons superiorly and the vertex metallic-violet (male), which is much reduced in
      the female; female with the vulvar lamina bilobed, lobes digitate, subequal in
      length to the interval separating their tips...*marcella*

1'   Rspl forming a loop with IR3, enclosing 3-4 cells in the forewings, 4-5 cells in
      the hind wings; hind wings with a basal brown area, reaching to or into the
      triangle (Fig. Li75), not extending to posterior wing margin; frons yellowish or
      greenish; female with the vulvar lamina extremely short, widely and shallowly
      emarginated...*simplex*

After: CALVERT (1908)

Micrathyria

Plate XXI, p. 115

Small tropical dragonflies which are mostly black with pale green markings, including a striped thorax and spotted abdomen. Abdominal segment 10 is very short, so that the large pale spot on segment 7 superficially seems to be on segment 8 (DUNKLE, 1989). They inhabit ponds of different types, ranging from highly vegetated and sunny ones to shaded forest ponds with little vegetation (DONNELLY, 1992).

1	2 cells between midrib of anal loop and triangle (Fig. Li76)	2
1'	1 cell between midrib of anal loop and triangle (Fig. Li77)	9
2	Forewing arculus distal to second antenodal crossvein	3
2'	Forewing arculus proximal to second antenodal crossvein	4

3   No throughgoing cell between MA and CuP (discoidal field) of hindwing; pale yellow band on mesepimeron as broad as two dark bands and intervening pale band below it, parallel sided .... *laevigata*

3'   At least one throughgoing cell between MA and CuP (discoidal field) of hindwing .... *dictynna*

4   Forewing triangle crossed 5

4'   Forewing triangle free 6

5   No throughgoing cell between MA and CuP (discoidal field) of hindwing; thorax color pattern as in Fig. Li91; male anal appendages as in Fig. Li79 ...... *didyma*

5'   At least one throughgoing cell between MA and CuP (discoidal field) of hindwing; male anal appendages as in Fig. Li78 .... *atra*

6   Forewing subtriangle 2- or 3-celled 7

6'   Forewing subtriangle 1-celled .... *catenata*

7   Superior appendages in dorsal view with their tips diverging, in profile showing no denticles, but the apical two-thirds very convex inferiorly, an external row of 5-6 denticles on the middle third, visible in dorsal or ventral view (Fig. Li80); brown metepimeral stripe running off from about midheight of the second lateral thoracic suture; abdominal segments 5 and 6 with no pale markings; anterior lamina and genital lobe subequally prominent .... *dissocians*

7'   Superior appendages in dorsal view not diverging, in profile showing teeth or denticles 8

8   Brown metepimeral stripe absent (Fig. Li89); abdominal segments 5 and 6 with pale stripes or spots; superior appendages in dorsal view with their tips not diverging, in profile with an almost straight inferior row of 6-8 denticles on the middle third (Fig. Li81); anterior lamina less prominent than the adjacent parts; vulvar lamina reaching to one third of the lateral margin of segment 9, not attaining the base of the style-like processes, its apical margin with a median notch .... *hagenii*

8'   Brown metepimeral stripe present, running off from near the lower end of the second lateral thoracic suture (Fig. Li90); abdominal segments 5 and 6 with no, or very small basal, pale marks; superior appendages in dorsal view converging, in profile with two inferior teeth at two-fifths and three-fifths their length respectively and separated from each other by a concavity, an external row of

Plate XXI

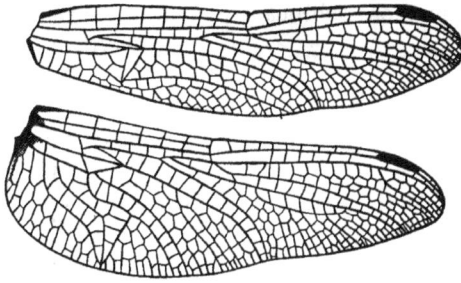
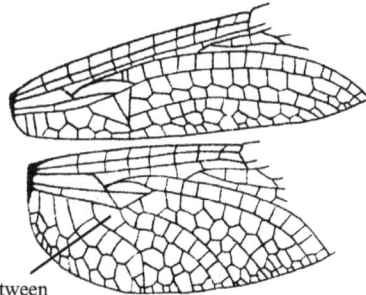

Fig. Li76: *Micrathyria hagenii*

cell between
midrip of anal loop and triangle

Fig. Li77: *Micrathyria caerulistyla*

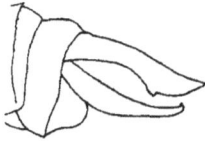
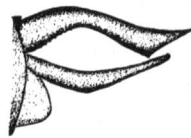
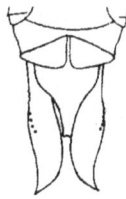

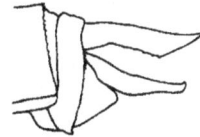

Li78: *M. atra*   **Li79**: *M. didyma*

Li81: *M. hagenii*

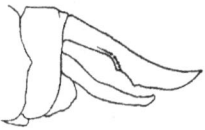

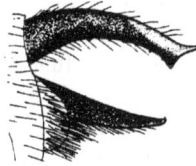

Li82: *M. ocellata*   **Li83**: *M. mengeri*

Li80: *M. dissocians*

Figs. Li78-85: *Micrathyria*, male abdominal
appenages, lateral view (except for *M. dissocians*,
left figure, which is dorsal view)

Li84: *M. pseudeximia*

Li85: *M. caerulistyla*

Li86: *M. atra*   **Li87**: *M. caerulistyla*   **Li88**: *M. debilis*

Li89: *M. hagenii*   **Li90**: *M. ocellata*   **Li91**: *M. didyma*

Figs. Li86-91: *Micrathyria*, thoracic color pattern

denticles running proximad from the distal tooth, but external to the proximal tooth (Fig. Li82); anterior lamina and genital lobe subequally prominent; vulvar lamina almost as in *hagenii* .... *ocellata*

9   Forewing subtriangle 1-celled **12**

9'   Forewing subtriangle 2- or 3-celled **10**

10   Forewing subtriangle 3-celled .... *schumanni*

10'   Forewing subtriangle 2-celled **11**

11   Thorax almost completely dark in male; large rounded pale lateral spots on abdominal segment 7 in both sexes; male superior appendages decurved in lateral view, tips convergent in dorsal view .... *aequalis*

11'   Thorax with small pale markings in male; abdominal segment 7 in both sexes with thin yellow line on side; male superior appendage straight in lateral view (Fig. Li83), tips divergent in dorsal view .... *mengeri*

12   Tips of male superior appendage dark; femora basally pale .... *tibialis*

12'   Tips of male superior appendage with pale markings; femora basally dark **13**

13   Male superior appendage strongly decurved in lateral view, nearly forming a rounded right angle, obscurely greenish grey at tip, expanded with venter convex downward (Fig. Li84); forewing arculus close to second antenodal crossvein .... *pseudeximia*

13'   Male superior appendage evenly decurved in lateral view **14**

14   Male superior appendages blue on dorsum of apical half, tip expanded with ventral concavity separating a sharp tip and a sharp sub-apical projection (Fig. Li85); 7 or 8 antenodals in hind wing; forewing arculus close to mid-point between first and second antenodal crossvein (Fig. Li77); thoracic color pattern as in Fig. Li87 .... *caerulistyla*

14'   Male superior appendages not as above; 6 antenodals in hind wing; thorax as in Fig. Li88 .... *debilis*

Compiled from: ALAYO (1968), CALVERT (1908, 1909), DONNELLY (1992), DUNKLE (1995), RIS (1909-1919), SANTOS (1949, 1954), WESTFALL (1992)

Nephepeltia

Small sized dragonflies (abdomen less than 20 mm), like a small *Micrathyria. N. phryne* is known to prefer small vegetated streams and stream pools, where males typically perch on stems along the waters edge with abdomen raised above the horizontal (MICHALSKI, 1988).

1   Two crossveins behind pterostigma, male without a spine on thoracic sternum .... *leonardina*

1'   One crossvein behind pterostigma, male with a definite conical bump or spine on thoracic sternum behind hind legs **2**

2   One cell between forewing subtriangle and posterior edge of wing, male bump wider than high .... *flavifrons*

2'   Two cells between forewing subtriangle and posterior edge of wing, male spine
       at least as high as wide_____**3**
3   Male spine pointed and twice as tall as wide_____***phryne***
3'   Male spine blunt and as tall as wide_____***chalconata***

After: DUNKLE (in litt.)

Oligoclada
Plate XXII, p. 121

Small sized dragonflies (abdomen 16-19 mm). Thorax of males bluish or metallic blue
(in *umbricola*), pruinose in mature individuals. They are characterized by a swollen
margin to the occiput, resembling in this detail the Cordullidae.
In South America, both species may occur at the same locality, but are separated
distinctly by their habitat preferences. *O. heliophila* is found in open, sunny areas like
clearing, wheras adults of *O. umbricola* are found along well shaded streams in the
forest. They perch flat on the bank or leaves with abdomen and wings in same plane,
and are fast flying and remarkably wary (BORROR, 1931).

1   Posterior margin of occiput with a short finger-like projection on either side (Fig.
       Li94); tooth of hamule on anterior side and directed caudad; 6 or more large
       teeth on ventral surface of superior appendage (Fig. Li92)_____***heliophila***
1'   Swollen posterior margin of occiput without such projections (Fig. Li95); tooth
       of hamule on posterior side and directed cephalad; 2-4 small teeth on ventral
       surface of superior appendage (Fig. Li93)_____***umbricola***

After: BORROR (1931)

Orthemis
Plate XXII, p. 121

Medium to large sized dragonflies, some rather stout bodied. Males are generally red
coloured with or without distinct black markings on the abdominal segments. Females
with lateral flaps on segment 8. They inhabit a variety of habitats, ranging from highly
artificial man-made ponds (*O. ferruginea*) to small streams and swamps in dense
primary forest (*O. cultriformis*).

Males
1   Hind wings with four or more rows of cells between A2 and the hind margin of
       the wing at the level of the hind angle of triangle (Fig. Li97)_____**2**
1'   Hind wings with three rows of cells between A2 and the hind margin of the wing
       at level of the hind angle of triangle_____**5**
2   Labium with a median dark band present_____**4**

2'   Labium with no median black band (its place sometimes taken by pale brown);
        femora and tibiae superiorly pale brown, tips of wings clear or slightly smoky;
        abdomen of mature male purple or bright red, thorax darkening with age.........**3**

3   Frons of male usually violet; tenerals with color pattern on sides of thorax
        consisting of distinct whitish lines on dark ground; a black mark behind base of
        hind leg present_____*ferruginea*

3'   Frons of male usually red; tenerals at most with only an obscure color pattern on
        sides of thorax, no black mark behind base of hind leg_____*discolor*

4   Femora and tibiae dark brown above, almost black on tibiae; tips of wings to
        distal end of pterostigma dark brown; brown on thorax dominated by the
        intervening yellow_____*ferruginea sulphurata*

4'   Femora and tibiae superiorly reddish brown; tips of wings clear or smoky; sides
        of thorax brown, with two horizontal longitudinal greenish or greenish-yellow
        bands, the upper beginning on the upper end of the mesinfraepisternum and
        lower end of the mesepisternum and extending to the upper half of the hind
        margin of the metepimeron, slightly interrupted at the humeral and second lateral
        sutures; from this band a short indistinct stripe runs upward bordering the
        humeral suture anteriorly; the lower band borders the latero-ventral metathoracic
        carina for its whole length, and is continued forward to the base of the second leg
        by two separated larger spots; (male unknown)_____*aequilibris*

(4''   Clypeus above blue; face reddish in the middle, shading into yellowish above
        and black below, the sides broadly yellow; legs blackish, femora pale beneath;
        wings hyaline with brown at the tips and brown pterostigma; thorax and
        abdomen reddish above, thorax with a yellow antehumeral stripe and a yellow
        stripe along the middorsal carina; a broader dorsal yellow stripe between the
        wings, narrowly continued on the abdomen for nearly its whole length
        _____*flavopicta*)

5   Arculus at, or proximal to, the second antenodal on the hind wings; narrow pale
        stripes on sides of thorax, that of the mesepimeron 0.5 mm wide; abdomen
        bright red; anterior lamina projecting not half as far as hamule; hamule like that
        of ferruginea, but larger and heavier, apex of outer branch less produced
        _____*biolleyi*

5'   Arculus distal to the second antenodal on the hind wing and, usually, also on the
        forewings; dorsum of abdominal segments 4-7 chiefly reddish or luteous with at
        most only a narrow darker stripe each side_____**6**

6   Genital hamule of male similar to that of *ferruginea*, but its outer branch less
        produced, anterior lamina projecting almost as much as the hamule (Fig. Li96)
        _____*levis*

6'   Genital hamule of male more prominent than in any other species of the genus,
        more prominent than anterior lamina or genital lobe; its outer branch longer than
        the inner, rolled inward at its extremity, touching its fellow of the opposite side
        and forming a covering below the shorter, hook-like, inner branch; extreme
        apices of both branches spinelike and directed forward_____*cultriformis*

Compiled from: CALVERT (1908, 1909), DUNKLE (in litt.), KIRBY (1889), MICHALSKI
(1988)

Pachydiplax

Pachydiplax longipennis

A small to medium sized species. Mature males are characterized by a white face, metallic green eyes, black and yellow striped thorax, and a pale blue tapered abdomen with a black tip. Hindwings with two black streaks within an amber spot at the base. In juveniles, the abdomen is black with two interrupted yellow stripes each side. The species inhabits almost any type of still water, including ponds, marshes, bays, ditches, and swamps. (DUNKLE, 1989)

Paltothemis

Paltothemis lineatipes

This is a medium sized dragonfly (abdomen about 50 mm) with a conspicuous red abdomen, occupying rocky hill streams in brushlands and forests. Males perch on rocks in the stream and occasionally make short patrolling flights low over water (DUNKLE, 1978).

Pantala

These „....are medium sized dragonflies with a streamlined tear-drop shaped body that tapers from the large round head to the tip of the pointed abdomen. Their hindwings are triangular, broad at the base, allowing them to remain aloft for hours." (DUNKLE, 1989).

1   Body yellowish, hind wings yellowish between the anal margin and A2, but no brown spot there, wings often a little brownish at tips; superior appendages of male 3.5 mm long, inferior appendage two-thirds as long................*flavescens*

1'   Body brownish, hind wings with a brown spot between anal margin and A2; superior appendage of the male 3 mm long, inferior appendage three-fourths as long..*hymenaea*

After: CALVERT (1908)

Perithemis

„Dragonflies of this genus are readily recognized by their small size and chunky build, plus orange wings of the male. The abdomen is spindle-shaped, tapered at both ends."(DUNKLE, 1989). As most species are highly sexual dimorphic, it is helpful for identification to have both male and female for examination.

1   At least one of triangles and subtriangles crossed..**2**
1'   All triangles and subtriangles free, sides of thorax light ochraceous brown, unmarked or only diffusely clouded above the metastigma; male wings golden yellow, venation and pterostigma red, very exceptionally some dark spots at the triangles; female highly polymorphous, wings hyaline with golden yellow, golden yellow and brown, or brown to blackish markings, proximal ones at the triangles, distal ones mostly developed as transverse bands with the middle of the band or spot at the nodus or very slightly distal or distinctly proximal thereto; proximal and distal bands often confluent, in the costal stripe alone, or more broadly, or even entirely..*mooma*
2   Legs dark brown, except the flexor side of first femora and the extensor side of all tibiae, which are light ochraceous yellow; thoracic dorsum purplish brown with slightly diffuse greenish antehumeral stripes, sides dull olive green with diffuse, often incomplete brownish bands at metastigma and posterior lateral suture; abdomen each side with a complete brownish-black longitudinal band from segment 4-9; male wings light and somewhat dull yellow with strongly contrasting dark red venation or richer golden yellow with venation equally contrasting, often a diffuse clearing between the triangular and nodal regions, very rarely a trace of dark spots in the triangular region where as a rule only the yellow ground color is somewhat deepened; pterostigma deep dark red; female polymorphous: yellow markings alone, always including the wing bases, or dark brown elements within such yellow, often a yellow to blackish ray in costal field to half-way between nodus and pterostigma, mostly the apex of hind wing brown ....*domitia*
2'   Legs light ochraceous yellow, unicolorous; thorax very light golden brown, sides slightly greenish, no dark markings; abdomen with basal segments almost unmarked, segments 7-9 each side with a narrow oblique dark stripe; male wings rich and pure golden yellow, venation the same color as the pterostigma or, in extreme cases, to the apex, this cleared apical area only showing blackish venation, no dark wing markings; female isochromatic, color slightly duller than in males, less dull distal to the triangle..*electra*

After: Ris (1930)

Planiplax

Males of *P. phoenicura* resemble very closely the males of *Erythemis peruviana*, being black on head, legs and thorax, and bright, clear red on the whole of the abdomen. Apart from the characters given in the key to the genera, the two may be distinguished in the field as follows: males of *E. peruviana* have segments 1 and 2 of the abdomen quite black; in *P. phoenicura*, all of the abdomen is red, including the male appendages

Plate XXII

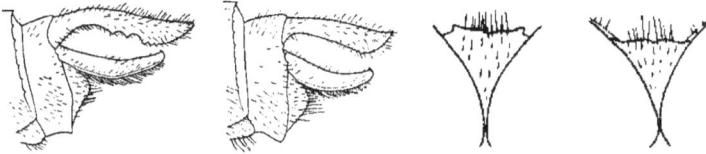

Li92: *O. heliophila*   Li93: *O. umbricola*   Li94: *O. heliophila*   Li95: *O. umbricola*

Fig. Li92 & 93: *Oligoclada*, male anal appendages, lateral view

Fig. Li94 & 95: *Oligoclada*, dorsal view of occiput

Fig. Li96: *Orthemis levis*, male accessory genitalia, lateral view

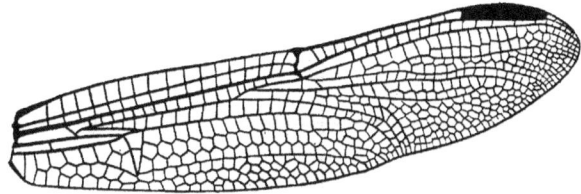

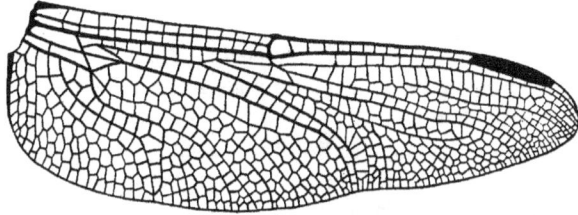

Li98   Li99

Figs. Li98 & 99: *Sympetrum*, ventral view of male abdominal appendages
Li98: *S. illotum*; **Li99**: *S. nigrocreatum*

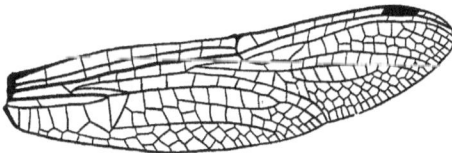

Fig. Li97: *Orthemis ferruginea*

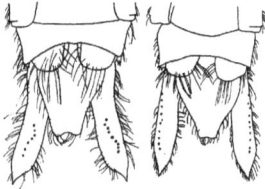

Fig. Li100: *Idiataphe cubensis*, front wing

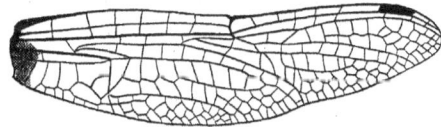

Fig. Li101: *Macrodiplax balteata*, front wing

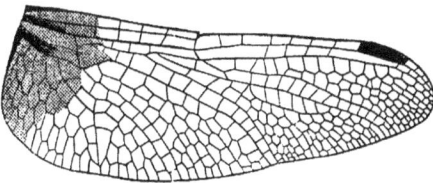

Fig. Li102: *Pachydiplax longipennis*, hind wing

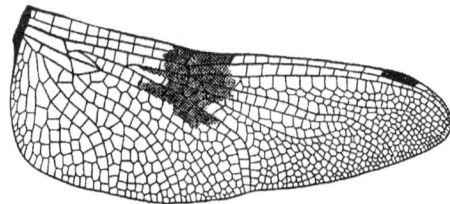

Fig. Li103: *Tholymis citrina*, hind wing

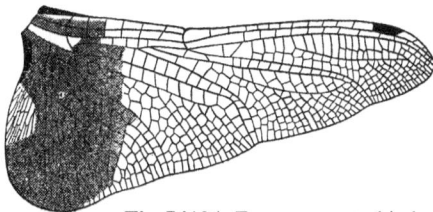

Fig. Li104: *Tramea onusta*, hind wing

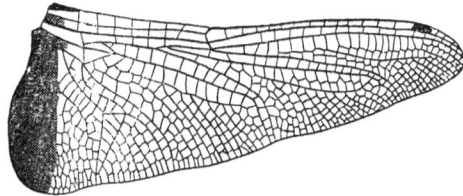

Fig. Li105: *Tramea abdominalis*, hind wing

(MICHALSKI, 1988). Males usually perch out from shore on twig tips at ponds (DUNKLE, in litt.).

1   Forewing triangle and subtriangle each 1-celled, 2 rows of cells beyond forewing
      triangle_____*phoenicura*
1'   Forewing triangle 2-celled, forewing subtriangle 3-celled, 3 rows of cells beyond
      forewing triangle_____*sanguiniventris*

After: DUNKLE (in litt.)

Pseudoleon

Pseudoleon superbus

Medium sized dragonfly (abdomen 42 mm), very strikingly colored with a heavy pattern of brown upon the wings. Eyes with about 8 alternating light and dark stripes of equal width extending vertically across their surface. Abdomen brownish with paired oblique pale streaks forming a series of V's opening backward (NEEDHAM, 1929). The species breeds in rocky streams, where it nearly always perches on rocks, on hot days raising both wings and abdomen (DUNKLE, in litt.).

Rhodopygia

Males of these medium sized dragonflies are easily recognized by being bright red all over except thorax above. They are found at somewhat sheltered pools and ditches (BELLE, 1998), where males perch along the water's edge on vertical stems and sticks.

1   Usually a single row of cells between Rs and Rspl (at most with one or two
      doubled cells), hind wings with brown-yellow basal spot at least extending to
      triangle_____*cardinalis*
1'   Usually two rows of cells between Rs and Rspl, wings hyaline_____*hinei*

After: BELLE (1998)

Sympetrum
Plate XXII, p. 121

Small to medium sized dragonflies which generally are found in swamps and on ponds or other still waters in open areas (CALVERT & CALVERT, 1917; DUNKLE, 1989). Only one species, *S. illotum*, occurs widely through Central America. *S. nigrocreatum* is restricted to higher altitudes of a Costa Rican mountain range (CALVERT, 1920a), and the North American *S. corruptum* only ranges south to Honduras and Belize.

1   Abdomen mostly greyish in mature males; legs black, femora and tibiae with a superior yellow stripe; area between IR3 and Rspl with two rows of cells; antenodals 7 on the front, 5 on the hind wings; tip of vertex truncated ....***corruptum***

1'   Abdomen bright red in mature males; legs luteous, wings with at least one dark basal streak, area between IR3 and Rspl with one row of cells; antenodals 8-10 on the front, 6 on the hind wing; tip of vertex emarginated **2**

2   Superior appendages of male with the inferior denticles, viewed from below, forming an almost straight line (Fig. Li99)..***illotum***

2'   Superior appendages of male with the inferior denticles, viewed from below, forming a reversed curve (Fig. Li98)..***nigrocreatum***

Compiled from: CALVERT (1908), CALVERT (1920)

Tauriphila

The species of this genus are similar to *Tramea* but are smaller and have shorter terminal abdominal appendages as well as a different arrangement of wing veins (DUNKLE, 1989). Males patrol over floating vegetation. Away from water they feed in sustained flight.

1   Hind wings with two posttriangular rows for a distance of 2-3 cells, then increasing to three rows; all wings with usually only one row of cells between IR3 and Rspl...***australis***

1'   Hind wings with two (occasionally three) post-triangular cells followed immediately by three rows; all wings with more often two rows of cells in the middle of the field between IR3 and Rspl..**2**

2   Abdominal segments 3-6 with a transverse apical black band occupying a fifth or a sixth of the length of the segments, confluent with a mid-dorsal black line which is triangularly dilated at the bases of 5-7 and is widened into a band on 8 or 9; superior appendages of male in dorsal view subparallel, in profile view upper edge almost straight, denticles of lower edge beginning at two-fifths' and ending at four-fifths' length of the appendage........................***azteca***

2'   Abdominal segments with no transverse or mid-dorsal bands; superior appendages of males in dorsal view converging, in profile view proximal two-thirds of upper edge strongly convex, terminal third nearly straight, denticles of lower edge beginning at two-fifths' and ending at two-thirds' length of the appendage..***argo***

After: CALVERT (1908)

Tholymis

Tholymis citrina

Medium sized libellulid with a typical yellowish to brownish spot at the nodus in hind wings (Fig. Li103). Entire body including the very large eyes of brownish coloration. This is one of the few libellulids which is crepuscular in its habits. Males are found patrolling low over ground just before dawn (MAY, 1979).

Tramea
Plate XXII, p. 121

„These medium sized red, brown, or black dragonflies have a streamlined tear-drop shaped body and broad hindwings." They „...have a narrow to wide dark crossband at the base of each hindwing." They „...feed in a sustained gliding flight, occasionally perching horizontally on the tip of a stem or twig." (DUNKLE, 1989)

1	Hind wings with dark basal colouring extending outward to the cubito-anal crossvein, the origin of Aspl, or, less often, to the proximal side of the triangle, nowhere extending beyond the level of the triangle (Fig. Li105); only one row of cells between Aspl and A2	**2**
1'	Hind wings with dark basal colouring extending outward into the triangle at least, its proximal margin very much broken or indented (Fig. Li104); two or more rows of cells between Aspl and A2 in at least part of their course	***onusta***
2	Sides of thorax with two broad oblique yellow bands	***calverti***
2'	Sides of thorax without yellow bands	**3**
3	Frons superiorly metallic blue	**4**
3'	Frons superiorly and vertex shining red or luteous, with no metallic-blue; body colored bright red; basal spot on hind wing reddish-brown	***abdominalis***
4	Abdomen entirely grey to blackish in mature males; spot on hind wing black	***binotata***
4'	Abdomen red in mature males, with black on dorsum of segments 8 and 9; spot on hind wing brown	***insularis***

Derived from: CALVERT (1908), DE MARMELS & RACENIS (1982)

Uracis

Females of these medium sized dragonflies are easily recognized by their vulvar lamina, which projects backward beyond the tip of the abdomen. Distal ends of wings in mature individuals usually with brown to blackish coloration of variable extension. Mature males with abdomen bluish pruinose above. They occur around temporary waters and swamps with muddy banks, where the females deposit their eggs by inserting the

LIBELLULIDAE   125

abdomen up to the 8th segment into the substrate while in flight, a behavior generally similar to that found in cordulegastrid dragonflies.

1   Supratriangles of forewings free; forewing with 1-3 cubital crossveins
        ....***imbuta***
1'   Supratriangles of fore and hind wing crossed; forewing with 3-7 cubital crossveins..**2**
2   Discoidal field of hind wing with two rows of cells, reaching or surpassing the level of bifurcation of R2+2 ....***fastigiata***
2'   Discoidal field of hind wing with one row of cells reaching or surpassing the level of bifurcation of R2+2; discoidal field in forewing with 3 cells, followed by two rows of cells, reaching the level of subnodus; two rows of cells in anal field
        ....***turrialba***

After: COSTA & SANTOS (1997)

Zenithoptera

Zenithoptera americana

These are rather small, blue-black dragonflies that have deep, reflective bluish-purple-black wings. „They resemble butterflies of the genus *Heliconius*, and go so far as to perch with their wings folded over their backs, as butterflies do." (MICHALSKI, 1988).

Acknowledgements

It has taken more than three years to turn the rough concept of this compilation into a readable version. During the time of preparation it was not always easy to keep myself motivated to go ahead with this challenging task. Therefore I'm especially grateful to those people who were never tired to contribute with critical comments, inspiring discussions, or literature.

I wish to thank FRANK SUHLING and ANDREAS MARTENS for their encouragement in every stage of my work. Without them I probably would never have come out with this project at all. Special thanks also to SID DUNKLE who helped to solve many problems in the design of several keys during the final stage of the project. I thank FRANK SUHLING for testing earlier versions of the manuscript in the field, and for giving helpul advices regarding the practical aspect of working with the keys on location.

For guiding my interest to Central America and for helping in every possible way during my research trips to Costa Rica, I'm very grateful to HEINZ HOFFMANN. My appreciation goes also to the LA SELVA BIOLOGICAL STATION crew, Costa Rica, for providing excellent research facilities among a fascinating environment. It would have been a poor time without the famous, delicious food they were constantly preparing there.

Glossary

abdominal appendages	male cerci and inferior appendages collectively
acuminate	ending in a long, tapering point
acute	acute-angled (markedly less than 90°) or sharp-pointed
angulate	angular, not rounded
annular	ringlike
antealar carina	cuticular ridge on the meso- and metathorax just anterior to the wing bases
antealar crest	triangular group of carinae on the mesepisternum at the upper end of the middorsal carina, just anterior to the base of the forewings
antealar sinus	depression just anterior to the base of forewings, outlined by the antealar crest
anterior	at the front; toward the front, relative to another feature
anterior hamules	hinged, platelike sclerites at the anterolateral corners of the genital fossa
anterior lamina	plate-like sclerite(s) forming anterior margins of the genital fossa
apex, apical	pertaining to the tip of wing, abdomen or segments of abdomen, remote from the thorax
attenuate	very thin, drawn out
auricle	spiracle, opening through which the air enters and leaves the body
bidentate	two-toothed
bifid	two-cleft, forked
bifurcation	structure that is two-forked, divided like a Y
billneate	two-lincd
bilobed	having two rounded divisions
brace-vein	crossvein slanted proximally from the posterobasal corner of the pterostigma to vein R2
bristle	conspicuously strong and stout hair
carina	narrow cuticular ridge
caudal	pertaining to the tail; toward the rear end
cercus	highly modified appendage of the tenth abdominal segment
cleft	narrowly U- or V-shaped
compressed	flattened from side-to-side
confluent	flowing or joined smoothly together
constricted	drawn together, narrowed
contiguous	touching or in contact
converge	to approach; opposed to diverge
convex	curved outward
crepuscular	active at twilight, especially dusk (more rarely dawn)
cuneiform	wedge-shaped, more or less triangular in shape
decurved	curved downwards
dentate	toothed
denticles	small teeth or toothlike projections

denticulate	minutely dentate
depressed	flattened dorsally (on the back)
dilated	widened, expanded
distal	away from the base, outward; opposed to proximal
distal tooth	toothlike spine shortly before the apex of male cerci
diurnal	active during the day
divaricate	widely diverging, branching off at a large angle and spreading apart
diverge	to spread apart; bcome different from; opposed to converge
dorsal	pertaining to the back or upper side; opposed to ventral
elevated	raised above the surroundings
emarginate	notched
epiproct	dorsal cuticular flap or lobe projecting from the 10th abdominal segment, usually small or minute in adults
excised	cut out
ferruginous	rust-colored
flavescent	somewhat yellow
forcipate	forceps-like, curving inward and nearly meeting at the tips
fossa, genital	cavity, on the ventral surface of abdominal segment two of the male, into which the penis and genital hamules are recessed
fulvous	tawny
furcate	forked
fuscous	dark brown, approaching black
genital fossa	see fossa, genital
genital hamule	a pair of male sexual appendages found on abdominal segment two, one on either side of the penis
genital lobe	plate-like, ventral extensions of abdominal segment two in males of the Libellulidae and Corduliidae, which arise to either side of the genital fossa
hamular process	genital hamule
hyaline	clear, not obscured by color
incised	cut in a slit or narrow groove
inferior	lower, located beneath
interalar	between the wings
intercalated vein	longitudinal vein extending inward for a short distance from the distal wing margin between the main longitudinal veins
intersternite	transverse sclerite between the pro- and mesosternum, often extending upward so it is visible laterally between the pleura
laterally	to the sides
lentic	still water
lobed	having rounded divisions
lotic	flowing water
luteous	egg yellow
medial	towards the midline; opposed to lateral
mesal	toward the latidudinal centre

obscure	darkened; difficult to distinguish
paraproct	ventrolateral lobe or process projecting from the 10th abdominal segment; in males forming the inferior caudal appendages used to clasp the female (in Zygoptera)
pedunculate	havin a short, thick stalk
penultimate	state immediately preceding the last (ultimate) state
petiolate	having a narrow, stem-like or handle-like base
petiole	narrow, stem-like base, especially of the wings
posterior	at the rear; toward the rear, relative to another feature
produced	drawn out or prolonged
prominence	structure or area raised above or protruding from a surrounding area or adjacent portions of a margin
pronotum	broad, sclerotized, dorsal surface of the prothorax
propleuron	lateral surface of the prothorax
proximal	toward the base, inward; opposed to distal
pruinose	covered with a bluish-white bloom or exudate
pseudostigma	anterodistal area of wing, near where a true pterostigma would be found, but traversed by veins, often also more extensive and less discretely defined than a pterostigma
pubescent	clothed with short fine hair or down
recurved	curved backward
rufescent	somewhat reddish
rufous	reddish
sanguineous	blood red
scythe-like	bent to form a semicircle like a scythe
serrate	toothed or notched along the edge, like a saw
serrulate	minutely serate
seta	cuticular hairs or bristles produced by single, specialized epidermal cells; includes most of the small hairs of the body surface
spine	sharp cuticular projection that is fused with surrounding cuticle
spinulose	having small spines
spur	sharp cuticular projection that is movable at base
sternite	the underneath portion of the segment of the abdomen
subapical	just below or before the apex
subbasal	just beyond the base
subequal	approximately equal
subnodus	crossveins extending backward directly behind the nodus
subtriangular	roughly triangular in shape
superior	upper, located above
surpassing	extending beyond
suture	a seam where chitinous plates are joined
tarsus	terminal segment of the leg, the feet
teneral	adult recently emerged from the larval stage and not fully sclerotized or colored
thoracic	pertaining to the thorax
tibia	the forth segment of the leg, between femur and tarsus
transverse	lying or extending across

trifid	three-cleft
trochanter	the second segment of the leg, between coxa and tibia
trough-shaped	widely U-shaped
truncate	cut off squarely
tubercle	small, rounded, knoblike process
tuft	group of hairs standing close together
tumid	swollen
ventral	pertaining to the belly or underside; opposed to dorsal
vermiculate	wormlike in shape
vertex	sclerite on the top of the head where the ocelli are placed
vestigial	small or imperfectly formed, usual non-functional remnant of an organ that was more fully developed in earlier stages of individual or evolutionary development
vulvar spine	sharp, caudal extension of the sternum of the eight abdominal segment in females of some damselflies

After: MICHALSKI (1988), ROWE (1987), WESTFALL & MAY (1996)

Figure sources

*... modification of original drawing

1,2,6,7	ASKEW (1988)
3,4	WESTFALL & MAY (1996)
5,8,9	DUMONT (1991)*
10	MUNZ (1919)*
11	ROWE (1987)*
AE1,2,5	ALAYO (1968)
AE3,4	CARVALHO (1992)
AE6,7	PAULSON (1994)
AE8	BELYSHEV & HARITONOV (1978)
AE9	DUMONT (1991)*
AE10, 11	FÖRSTER
C1,2,4,7	MUNZ (1919)
C3,5,8,10	WESTFALL & MAY (1996)
C6,9	ALAYO (1968)
C11,29-33	WESTFALL & MAY (1996)*
C12	WESTFALL & MAY (1996)*, CALVERT (1908)
C13	DONNELLY (1967)
C14-16,18-21	LEONHARD (1977)
C17	GARRISON (1985)
C22,23	LEONHARD (1937)
C24,25	CALVERT (1908), GARRISON (1996)
C26	DUNKLE (1990)
C27,28	DONNELLY (1968)
C34	BICK & BICK (1995)
Ca1-41	GARRISON (1990)
F1-3, 6-12	ESQUIVEL (1991)*
F4,5,9	MUNZ (1919)*
G1	BELLE (1970)
G2,11	BELLE (1977)*
G3	BELLE & QUINTERO (1992)
G4	BELLE (1982)
G5	WESTFALL (1989)
G6	BELLE (1984b)*
G7,85-92	BELLE (1973)
G8	BELLE (1981)*
G9,68,69	BELLE (1988)
G10	BELLE (1984c)
G12-15	GARRISON (1986)
G16-18,27-29	KENNEDY (1946)
G19-22,35,36,70,74	BROOKS (1989)
G23-26	CALVERT (1920b)*
G30-33	BELLE (1994)
G36-38	BELLE (1993)
G39,40,46	BELLE (1989)

G41,42	DONNELLY (1986)
G43-45	BELLE (1980)
G47-67	GARRISON (1994)
G71-73,75-77, 78-84	DONNELLY (1979)
Le1&2	MUNZ (1919)
Le3,4,5,7	CALVERT (1908)
Le6,10	MAY (1993)
Le8,9	KORMONDY (1959)
Li1-23	BORROR (1945)*
Li24,74,75,76,97,100-105	ALAYO (1968)
Li25,27,69,70,72,73	CALVERT (1898)*
Li26	MICHALSKI (1988)
Li28-31,77,85,87	DONNELLY (1992)*
Li32,33	WESTFALL (1988)
Li34-63	BORROR (1942)
Li64,65	GARRISON (1992)
Li66	DONNELLY (1984)
Li67,68	MAY (1998)
Li71	COSTA (1990)
Li78,80-83,96	CALVERT (1908)
Li79,91	DONNELLY (1970)
Li84	WESTFALL (1992)
Li86,88-90	CALVERT (1908)*
Li92-95	BORROR (1931)
Li98,99	CALVERT (1920a)
Me1-3	MUNZ (1919)
Me4	DONNELLY (1965b, 1992)
Me5,6,11,14-17	CALVERT (1924)
Me7	MAY (1989)
Me8,13	WESTFALL & CUMMING (1956)*
Me9	BROOKS (1989)
Me10,12	DONNELLY (1989)
NC1-5	MAY (1991)*
Pe1,2	WILLIAMSON & WILLIAMSON (1924)
PL1,5,6,9,12,14	DONNELLY (1992)
PL2,4,7,8,10,11,13,15	CALVERT (1931)
PL3	BROOKS (1989)
PR1-4	MUNZ (1919)
PR5,6	CALVERT (1908)
PR7,8	ESQUIVEL (1994)

References

ALAYO, D.P., 1968. Las libélulas de Cuba. Parte II. Torreia, N.S. 3

ALCOCK, J., 1989. The mating system of Brechmorhoga pertinax (Hagen): The evolution of brief patrolling bouts in a "territorial" dragonfly (Odonata: Libellulidae). J. Ins. Behav. 2: 49-62.

ASKEW, R.R., 1988. The Dragonflies of Europe. Harley Books, Colchester.

BELLE, J., 1970. Studies on South American Gomphidae (Odonata), with special reference to the species from Surinam. Studies fauna Surinam and other Guyanas 8: 29-60.

BELLE, J., 1972. On Diaphlebia Selys, 1854 from Central America (Odonata: Gomphidae). Odonatologica 1: 63-71.

BELLE, J., 1973. A revision of the New World genus Progomphus Selys, 1854 (Anisoptera: Gomphidae). Odonatologica 2: 191-308.

BELLE, J., 1977. Notes on Aphylla obscura (Kirby, 1899)(Anisoptera: Gomphidae). Odonatologica 6: 7-12.

BELLE, J., 1980. A new species of Epigomphus from Guatemala (Odonata: Gomphidae). Ent. Ber. Amst. 40: 136-138.

BELLE, J., 1981. A new species of Phyllogomphoides from Ecuador (Odonata: Gomphidae). Ent. Ber. Amst. 41: 173-176.

BELLE, J., 1982. A review of the genus Archaeogomphus Williamson (Odonata, Gomphidae). Tijdschr. Ent. 125: 37-56.

BELLE, J., 1984a. Phyllogomphoides litoralis, a new species from Panama (Odonata: Gomphidae). Ent. Ber. Amst. 44: 174-175.

BELLE, J., 1984b. Progomphus maculatus, a new species from Venezuela (Odonata: Gomphidae). Ent. Ber. Amst. 44: 185-186.

BELLE, J., 1984c. A synopsis of the South American species of Phyllogomphoides, with a key and descriptions of three new taxa (Odonata, Gomphidae). Tijdschr.Ent. 127: 79-100.

BELLE, J., 1988. A synopsis of the species of Phyllocycla Calvert, with descriptions of four new taxa and a key to the genera of neotropical Gomphidae (Odonata, Gomphidae). Tijdschr. Ent. 131: 73-103.

BELLE, J., 1989a. A revision of the New World genus Neuraeshna Hagen, 1867 (Odonata: Aeshnidae). Tijdschr. Ent. 132: 259-284.

BELLE, J., 1989b. Epigomphus corniculatus, a new dragonfly from Costa Rica (Odonata: Gomphidae). Tijdschr. Ent. 132: 158-160.

BELLE, J., 1993. Epigomphus jannyae spec. nov., a new dragonfly from Panama (Anisoptera: Gomphidae). Odonatologica 22: 187-189.

BELLE, J., 1994. Three new neotropical Gomphidae from the genera Archaeogomphus Williamson, Cyanogomphus Selys and Epigomphus Hagen (Anisoptera). Odonatologica 23: 45-50.

BELLE, J., 1998. Synopsis of the Neotropical genus Rhodopygia Kirby, 1889 (Odonata: Libellulidae). Zool. Med. Leiden 72(1): 1-13.

BELLE, J. & D. QUINTERO, 1992. Clubtail dragonflies of Panama (Odonata: Anisoptera: Gomphidae). In: Quintero, D.A., A. (ed.). Insects of Panama and Mesoamerica. Selected Studies. Oxford University Press, Oxford: 91-101

BELYSHEV, B.F. & A.Y. HARITONOV, 1978. Determiner of dragonflies (genera of boreal faunistic kingdom and some contiguous territories, species of the USSR fauna). Nauka, Novosibirsk.

BICK, G.H. & J.C. BICK, 1988. A review of the males of the genus Philogenia, with descriptions of five new species from South America (Zygoptera: Megapodagrionidae). Odonatologica 17: 9-32.

BICK, G.H. & J.C. BICK, 1990. A revision of the neotropical genus Cora Selys, 1853 (Zygoptera: Polythoridae). Odonatologica 19: 117-143.

BICK, G.H. & J.C. BICK, 1995. A review of the genus Telebasis with descriptions of eight new species (Zygoptera: Coenagrionidae). Odonatologica 24: 11-44.

BICK, G.H. & J.C. BICK, 1996. Females of the genus Telebasis, with a description of T. bastiaani spec. nov. from Venezuela (Zygoptera: Coenagrionidae). Odonatologica 25: 1-15.

BORROR, D.J., 1931. The genus Oligoclada (Odonata). Misc. Publ. Mus. Zool. Univ. Mich. 23: 1-42.

BORROR, D.J., 1942. A revision of the libelluline genus Erythrodiplax (Odonata). Ohio Ste. Univ., Colombus

BORROR, D.J., 1945. A key to the new world genera of Libellulidae (Odonata). Ann. Ent. Soc. Am. 38:

BROOKS, S.J., 1989. New dragonflies (Odonata) from Costa Rica. Tijdschr. Ent. 132: 163-176.

CALVERT, P.P., 1898. The Odonate genus Macrothemis and its Allies. Proc. Boston Soc. Nat. Hist. 28: 301-332.

CALVERT, P.P., 1903. Synopsis of three species of of Coryphaeschna. Ent. News 14: 8-10.

CALVERT, P.P., 1908. Neuroptera, Fam. Odonata. In: Porter, R.H. (ed.). Biologia-Centrali Americana. Dulau & Co., London: 420

CALVERT, P.P., 1909. Contributions to a knowledge of the Odonata of the Neotropical region, exclusive of Mexico and Central America. Ann. Carn. Mus. 6: 73-281.

CALVERT, P.P., 1917. Studies on Costa Rican Odonata. VIII. A new genus allied to Cora. Ent. News 28: 259-263.

CALVERT, P.P., 1920a. Studies on Costa Rican Odonata. IX. Sympetrum with description of a new species. Ent. News 31: 253-259.

CALVERT, P.P., 1920b. The Costa Rican species of Epigomphus and their mutual mating adaptions (Odonata). Trans. Am. Ent. Soc. 46: 323-354.

CALVERT, P.P., 1924. The generic characters and the species of Philogenia Selys (Odonata: Agrionidae). Trans. Am. Ent. Soc. 50: 1-56.

CALVERT, P.P., 1931. The generic characters and the species of Palaemnema (Odonata: Agrionidae). Trans. Am. Ent. Soc. 57: 1-111.

CALVERT, P.P., 1956. The Neotropical species of the "subgenus Aeschna" sensu selysii 1883 (Odonata). Mem. Am. Ent. Soc. 15: 1-251.

CALVERT, A.S.& P.P. CALVERT, 1917. A year of Costa Rican Natural History. Macmillan, New York.

CARVALHO, A.L., 1989. Description of the larva of Neuraeschna costalis (Burmeister), with notes on its biology, and a key to the genera of Brazilian Aeshnidae larvae (Anisoptera). Odonatologica 18(4): 325-332.

CARVALHO, A.L., 1992. Revalidation of the genus Remartinia Navás, 1911, with the description of a new species and a key to the genera of neotropical Aeshnidae (Anisoptera). Odonatologica 21: 289-298.

COSTA, J.M. & T.C. SANTOS, 1997. Intra- and interspecific variation in the genus Uracis Rambur, 1842, with a key to the known species (Anisoptera: Libelllulidae). Odonatologica 26: 1-7.

CUMMING, R.B., 1954. Notes on the genus Metaleptobasis with the description of a new species from Panama (Odonata: Coenagriidae). Fla Ent. 37: 23-32.

DE MARMELS, J., 1989. Odonata or dragonflies from Cerro de la Neblina and the adjacent lowlands between the Rio Baria, the Casiquiare and the Rio Negro (Venezuela). I. Adults. Boln. Acad. Cienc., Caracas 25: 11-78, 89-91.

DE MARMELS, J., 1992. Dragonflies (Odonata) from the Sierras of Tapirapeco and Unturan, in the extreme south of Venezuela. Acta biol. venez. 14: 57-78.

DE MARMELS, J & J. RACENIS, 1982. An analysis of the cophysa-group of Tramea Hagen, with descriptions of two new species (Anisoptera: Libellulidae). Odonatologica 11(2): 109-128.

DONNELLY, T.W., 1965a. A new species of Ischnura from Guatemala, with revisionary notes on related North and Central American damselflies (Odonata: Coenagrionidae). Fla Ent. 48: 57-63.

DONNELLY, T.W., 1965b. Heteragrion eboratum, a new species of damselfly from Guatemala (Odonata: Megapodagrionidae). Proc. ent. Soc. Wash. 67: 96-100.

DONNELLY, T.W., 1967. The discovery of Chrysobasis in Central America, with the description of a new species (Odonata: Coenagrionidae). Fla. Ent. 50: 47-52.

DONNELLY, T.W., 1968. A new species of Enallagma from Central America (Odonata: Coenagrionidae). Fla Ent. 51: 101-105.

DONNELLY, T.W., 1970. The Odonata of Dominica, British West Indies. Smithsonian Contributions to Zoology 37: 1-20.

DONNELLY, T.W., 1979. The genus Phyllogomphoides in Middle America (Anisoptera: Gomphidae). Odonatologica 8: 245-265.

DONNELLY, T.W., 1984. A new species of Macrothemis from Central America with notes on the distinction between Brechmorhoga and Macrothemis. Fla Ent. 67: 169-174.

DONNELLY, T.W., 1986. Epigomphus westfalli spec. nov., a new dragonfly from Nicaragua (Anisoptera: Gomphidae). Odonatologica 15: 37-41.

DONNELLY, T.W., 1989a. A new species of Philogenia from Honduras (Odonata: Megapodagrionidae). Fla Ent. 72: 425-428.

DONNELLY, T.W., 1989b. Protoneura sulfurata, a new species of damselfly from Costa Rica, with notes of the circum-carribbean species of the genus (Odonata: Protoneuridae). Fla Ent. 72: 436-441.

DONNELLY, T.W., 1992. The Odonata of Central Panama and their position in the neotropical odonate fauna, with a checklist, and descriptions of new species. In: Quintero, D.A., A. (ed.). Insects of Panama and Mesoamerica: selected studies. Oxford University Press, Oxford:

DONNELLY, T.W. & P. ALAYO D., 1966. A new genus and species of damselfly from Guatemala and Cuba (Odonata: Coenagrionidae). Fla Ent. 49: 107-114.

DUMONT, H.J., 1991. Insecta V - Odonata of the Levant. The Israel Academy of Sciences and Humanities, Jerusalem.

DUNKLE, S.W., 1978. Notes on adult behavior and emergence of Paltothemis lineatipes Karsch, 1890 (Anisoptera: Libellulidae). Odonatologica 7: 277-279.

DUNKLE, S.W., 1989. Dragonflies of the Florida Peninsula, Bermuda and the Bahamas. Scientific Publishers, Gainesville.

DUNKLE, S.W., 1990. Damselflies of Florida, Bermuda and the Bahamas. Scientific Publishers, Gainesville.

DUNKLE, S.W., 1995. Geographical variation in Micrathyria mengeri Ris, with a description of M. mengeri watsoni ssp. nov. (Anisoptera: Libellulidae). Odonatologica 24: 45-51.

ESQUIVEL, C., 1991. Clave para identificar las familias de libelulas (Insecta: Odonata) presentes en Mexico y America Central. Brenesia 34: 15-26.

ESQUIVEL, C., 1994. Psaironeura selvatica sp. nov. (Odonata: Protoneuridae), a new damselfly from Costa Rica. Rev. Biol. Trop. 41: 703-707.

FINCKE, O.M., 1984. Giant damselflies in a tropical forest: Reproductive biology of Megaloprepus coerulatus with notes on Mecistogaster (Zygoptera: Pseudostigmatidae). Adv. Odonatol. 2: 13-27.

FINCKE, O.M., 1992a. Behavioural ecology of the Giant Damselflies of Barro Colorado Island, Panama. In: Quintero, D.A., A. (ed.). Insects of Panama and Mesoamerica: selected studies. Oxford University Press, Oxford.

FINCKE, O.M., 1992b. Interspecific competition for tree holes: Consequences for mating systems and coexistence in neotropical damselflies. Am. Nat. 139(1): 80-101.

GARRISON, R.W., 1973. The female of Libellula (Holotania) gaigei Gloyd, 1938 (Anisoptera: Libellulidae). Odonatologica 2: 109-113.

GARRISON, R.W., 1982. Archilestes neblina, a new... Occ. Pap. Mus. Zool. Univ. Michigan 702: 1-12.

GARRISON, R.W., 1985. Acanthagrion speculum spec. nov., a new damselfly from Costa Rica (Zygoptera: Coenagrionidae). Odonatologica 14: 37-44.

GARRISON, R.W., 1986. The genus Aphylla in Mexico and Central America, with a description of a new species, Aphylla angustifolia (Odonata: Gomphidae). Ann. Ent. Soc. Am. 79: 938-944.

GARRISON, R.W., 1990. A synopsis of the genus Hetaerina with descriptions of four new species (Odonata: Calopterygidae). Trans. Am. Ent. Soc. 116(1): 175-259.

GARRISON, R.W., 1992. Libellula mariae spec. nov., a new dragonfly from Costa Rica (Anisoptera: Libellulidae). Odonatologica 21: 85-89.

GARRISON, R.W., 1994. A revision of the New World genus Erpetogomphus Hagen in Selys (Odonata: Gomphidae). Tijdschr. Ent. 137: 173-269.

GARRISON, R.W., 1996. A synopsis of the Argia fissa group, with descriptions of two new species, A. anceps sp. n. and A. westfalli sp. n. (Zygoptera: Coenagrionidae). Odonatologica 25: 31-47.

GEIJSKES, D.C., 1943. Notes on Odonata of Surinam. III. The genus Coryphaeschna, with descriptions of a new species and of the nymph of C. virens. Ent. News 54: 61-72.

GEIJSKES, D.C., 1959. The aeschnine genus Staurophlebia. Notes on Odonata of Surinam VII. Stud. Fauna Suriname and other Guyanas 3: 147-172.

GLOYD, L.K., 1938. A new species of the genus Libellula from Yukatan. Occ. Pap. Mus. Zool. Univ. Mich. 377: 1-4, pl. 1.

GONZÁLEZ SORIANO, E., 1991. A new species of Amphipteryx Selys, 1853 from Oaxaca, Mexico (Zygoptera: Amphipterygidae). Odonatologica 20: 465-470.

GONZÁLEZ SORIANO, E. & M. DEL PILLAR VILLEDA C., 1978. The first Mexican record of Perissolestes magdalenae (Williamson & Williamson) (Zygoptera: Perilestidae). Notul. Odonatol. 1(2): 22-23.

GONZÁLEZ SORIANO, E., R. NOVELO GUTIÉRREZ & M. VERDUGO GARZA, 1982. Reproductive behavior of Plaemnema desiderata Selys (Odonata: Platystictidae). Adv. Odonatol. 1: 55-62.

GONZÁLEZ SORIANO, E. & M. VERDUGO GARZA, 1982. Studies on neotropical Odonata: The adult behaviour of Heteragrion alienum Williamson (Odonata: Megapodagrionidae). Folia Entomol. Mex. 52: 3-15.

GONZÁLEZ SORIANO, E. & M. VERDUGO GARZA, 1984a. Estrategias reproductivas en algunas especies de zigopteros Neotropicales (Insecta, Odonata). Folia Entomol. Mex. 61: 93-103.

GONZÁLEZ SORIANO, E. & M. VERDUGO GARZA, 1984b. Estudios en odonatos Neotropicales II: Notas sobre el comportamiento reproductivo de Cora marina Selys (Odonata: Polythoridae). Folia Entomol. Mex. 62: 3-15.

KENNEDY, C.H., 1940. The Miocora like dragonflies from Ecuador, with notes on Cora, Miocora, Kalocora, Josocora and Stenocora (Odonata: Polythoridae). Proc. Ent. Soc. Am. 33: 406-436.

KENNEDY, C.H., 1941. Perilestinae in Ecuador and Peru: revisional notes and descriptions (Lestidae: Odonata). Ann. Ent. Soc. Am. 34: 658-688.

KENNEDY, C.H., 1946. Epigomphus subquadrices, a new dragonfly (Odonata: Gomphidae) from Panama, with notes on E. quadrices and Eugomphus n. subgen. Ann. Ent. Soc. Am. 39: 662-666.

KIRBY, W.F., 1889. A revision of the subfamily Libellulinae, with descriptions of new Genera and Species. Trans. zool. Soc. Lond. 12: 249-348.

LEONHARD, J.W., 1937. A new Anisagrion from Panama, with notes on related species (Odonata: Zygoptera). Occ. Pap. Mus. Zool. Univ. Mich. 354: 1-7.

LEONHARD, J.W., 1977. A revisionary study of the genus Acanthagrion (Odonata: Zygoptera). Misc. Publ. Mus. Zool. Univ. Mich. 153: 1-46, 153-173.

MACHADO, A.B.M., 1954. „Elga santosi" sp. n. e redescricao de „Elga leptostyla" Ris, 1911 (Odonata, Libellulidae). Rev. Brasil. Biol. 14(3): 303-312.

MAY, M.L., 1979. Lista preliminar de nombre y clave para identificar los Odonata (caballitos) de Isla de barro Colorado. Smithsonian Tropical Research Institute & La Editorial Universitaria, Panama.

MAY, M.L., 1989. The status of Philogenia leonora Westfall & Cumming (Odonata: Megapodagrionidae). Odonatologica 18: 95-97.

MAY, M.L., 1991. A review of the genus Neocordulia, with a description Mesocordulia subgen. nov. and of Neocordulia griphus spec. nov. from Central America, and a note on Lauromacromia (Odonata: Corduliidae). Folia Entomol. Mex. 82: 17-67.

MAY, M.L., 1993. Lestes secula, a new species of damselfly (Odonata: Zygoptera: Lestidae) from Panama. J. N.Y. ent. Soc. 101: 410-416.

MAY, M.L., 1998. Macrothemis fallax, a new species of dragonfly from Central America (Anisoptera: Libellulidae), with a key to the male Macrothemis. Int. J. Odonatol. 1(2): 137-153.

MICHALSKI, J., 1988. A catalogue and guide to the dragonflies of Trinidad (order Odonata). Univ. West Indies Dep. Zool. Occas. Pap. 16: i-ii, 1-146.

MONTGOMERY, B.E., 1967. Studies in the Polythoridae. 1. A synopsis of the family, with keys to genera and species, information on types, and the descriptionof a new species. Acta biol. venez. 5: 123-158.

MUNZ, P.A., 1919. A venational study of the suborder Zygoptera. Mem. Am. Ent. Soc. 3: 1-78.

NEEDHAM, J.G. & H.B. HEYWOOD, 1929. A Handbook of the Dragonflies of North America. Charles C. Thomas, Springfield.

NOVELO GUTIÉRREZ, R., 1989. The larva of Agriogomphus tumens Calvert (Anisoptera: Gomphidae). Odonatologica 18(2): 203-207.

NOVELO GUTIÉRREZ, R., 1995. La nayade de Brechmorhoga praecox (Hagen, 1861), y notas sobre las nayades de B. rapax Calvert, 1898, B. vivax Calvert, 1906 y B. mendax (Hagen, 1861)(Odonata: Libellullidae). Folia Entomol. Mex. 94: 33-40.

NOVELO GUTIÉRREZ, R. & J. PEÑA OLMEDO, 1989. The subspecies of Ischnura posita (Hagen, 1861), with description of I. p. atezca ssp. nov. (Zygoptera: Coenagrionidae). Odonatologica 18(1): 43-49.

PAULSON, D.R., 1982. Odonata. In: Hurlbert, S.H. & A. Villalobos Figueroa (ed.). Aquatic biota of Mexico, Central America and the West Indies. San Diego State University, San Diego: 249-277

PAULSON, D.R., 1994. Two new species of Coryphaeschna from Middle America, and a discussion of the red species of the genus (Anisoptera: Aeshnidae). Odonatologica 23: 379-398.

PAULSON, D.R., 1997. Odonata of Middle America, by country. URL: http://www.ups.edu/biology/museum/ODofMA.html

RAMÍREZ, A., 1992. Description and natural history of Costa Rican dragonfly larvae. 1. Heteragrion erythrogastrum Selys, 1886 (Zygoptera: Megapodagrionidae). Odonatologica 21(3): 361-365.

RAMÍREZ, A., 1994. Descrpcion e historia natural de las nayades de odonatos de Costa Rica. II: Archilestes neblina (Garrison, 1982) (Odonata, Lestidae), con una clave para las especies del genero en Costa Rica. Folia Entomol. Mex. 90: 9-16.

RAMÍREZ, A.; PAULSON, D.R. & C. ESQUIVEL, submitted. Odonata of Costa Rica. I: Diversity and checklist of species. Rev. Biol. Trop.

RIS, F., 1918. Libellen (Odonata) aus der Region der amerikanischen Kordilleren von Costarica bis Catamarca. Arch. Naturgesch. (A) 82: 1-197.

RIS, F., 1930. A revision of the Libelluline genus Perithemis (Odonata). Misc. Publ. Mus. Zool. Univ. Mich. 21: 1-50.

RIS, R., 1909-1919. Libellulinen monografisch bearbeitet. Collns. zool. Sélys-Longchamps 16: 1146-1147.

ROWE, R.J., 1987. The Dragonflies of New Zealand. Auckland University Press, Auckland.

SANTOS, N.D. DOS, 1949. Contribucao ao conhecimento da fauna de Pirassununga. 6. Descricao de femea de Micrathyria catenata Calv., 1909 e notas sobre M. ocellata dentiens Calvert, 1909 (Odonata: Libellulidae). Rev. Ent. 20: 159-164.

SANTOS, N.D. DOS, 1954. Revisao de Micrathyria didyma (Selys, 1857) e suas correlatas (Libellulidae, Odonata). Arq. Mus. Nac. (Rio de Janeiro) 42: 491-498.

SHELLY, T.E., 1982. Comparative foraging behavior of light- versus shade-seeking adult damselflies in a lowland neotropical forest (Odonata: Zygoptera). Physiol. Zool. 55(4): 335-343.

SIVA-JOTHY, M.T. & Y. TSUBAKI, 1989. Variation in copulation duration in Mnais pruinosa pruinosa Selys (Odonata: Calopterygidae). 1. Alternative mate-securing tactics and sperm precedence. Behav. Ecol. Sociobiol. 24: 39-45.

ST. QUENTIN, D., 1960. Zur Kenntnis der Agrioninae (Coenagrioninae) Südamerikas (Odonata). Beitr. neotrop. Fauna 2: 45-64.

WESTFALL, M.J., 1988. Elasmothemis gen. nov., a new genus related to Dythemis (Anisoptera: Libellulidae). Odonatologica 17: 419-428.

WESTFALL, M.J., 1989. The larvae of Desmogomphus paucinervis (Selys, 1873) and Perigomphus pallidistylus (Belle, 1972) (Anisoptera: Gomphidae). Odonatologica 18: 90-106.

WESTFALL, M.J., 1992. Notes on Micrathyria, with descriptions of M. pseudeximia sp.n., M. occipita sp.n., M. dunklei sp.n. and M. divergens sp.n. (Anisoptera: Libellulidae). Odonatologica 21: 203-218.

WESTFALL, M.J. & R.B. CUMMING, 1956. Two new species of Philogenia from the Panama Canal Zone (Odonata: Coenagriidae). Bull. Fla. St. Mus. biol. Sci. 1: 241-252.

WESTFALL, M.J. & M.L. MAY, 1996. Damselflies of North America. Scientific Publishers, Gainesville.

WILLIAMS, F.X., 1937. Notes on the biology og Gynacantha nervosa Rambur (Aeschninae), a crepuscular dragonfly in Guatemala. Pan-Pacific Ent. 13(1-2): 1-8.

WILLIAMSON, E.B., 1915. Notes on Neotropical dragonflies, or Odonata. Proc. U.S. Nat. Mus. 48: 601-638.

WILLIAMSON, E.B., 1917. The genus Neoneura (Odonata). Trans. Amer. Ent. Soc. 43: 211-246.

WILLIAMSON, E.B., 1919a. Results of the University of Michigan-Williamson Expedition to Colombia, 1916-1917. III. Archaeogomphus, a new genus of dragon-flies (Odonata). Occ. Pap. Mus. Zool. Univ. Mich. 63: 1-61.

WILLIAMSON, E.B., 1919b. Results of the University of Michigan-Williamson Expedition to Colombia, 1916-1917. IV. Notes on species of the genus Heteragrion Selys with descriptions of new species (Odonata). Occ. Pap. Mus. Zool. Univ. Michigan 68: 1-65.

WILLIAMSON, E.B., 1923a. Notes on American species of Triacanthagyna and Gynacantha. Misc. Publ. Mus. Zool. Univ. Mich. 9: 1-80.

WILLIAMSON, E.B., 1923b. Notes on the genus Erythemis with a description of a new species (Odonata). Misc. Publ. Mus.Zool. Univ. Mich. 13: 1-46.

WILLIAMSON, E.B., 1930. Two new neotropical Aeshnines (Odonata). Occ. Pap. Mus. Zool. Univ. Mich. 218: 1-15.

WILLIAMSON, E.B. & J.H. WILLIAMSON, 1924. The genus Perilestes (Odonata). Misc. Publ. Mus. Zool. Univ. Mich. 14: 1-36.

140

Index

www.ingramcontent.com/pod-product-compliance
Lightning Source LLC
Chambersburg PA
CBHW081540220326
41598CB00036B/6497